OFF-GRID SOLAR POWER MADE EASY

DESIGN AND INSTALLATION OF PHOTOVOLTAIC SYSTEM FOR RVS, VANS, CABINS, BOATS AND TINY HOMES

Table of Contents

Introduction

If you are reading this, it is because you are serious about going off the grid. I am well aware that there are plenty of sources of information out there for you to choose from. So, it is an honor to have you go through these pages.

Granted, not all sources of information are credible and reliable. So, great care has been taken in making this book as practical and accurate as possible. I continuously work on the improvement of this book to make it the best quality, among other solar power books.

The solar power system can not only help you get off the grid but also become completely self-reliant.

This is something the key here.

You see, most of the folks I talk to are concerned about having power, especially in those times when there are disruptions in the regular service. It is not so much a question of saving money, but it is a question of being self-reliant.

For who is this book?

This book is designed for beginners, for those who know a little to nothing about solar power and photovoltaics. If you are a professional electrician or photovoltaic practitioner this book might not be very useful for you.

Think about the many ways in which your power supply could be disrupted. Anything is possible. Even something as simple as a faulty line can wreak havoc on a community.

Furthermore, the cost of the average electric bill has gone up over the years. According to the US Bureau of Labor Statistics, the price of electricity has gone by 468% from 1930 to 2019.

468%!

With the prices of oil and natural gas, the price of electricity and heating is not going down any time soon. This is even worse when you consider that wages don't increase nearly as much as prices do. From 1973 to 2013, the hourly compensation of workers, on average, rose 9.2%.

Let's consider this situation:

A typical 5-kilowatt system can cost approximately $18,000 to $25,000. If you are not familiar with these costs, you might be shaking your head right now. But fear not, most states have a rebate plan in which you can deduct taxes from the cost of your solar panel system.

Yes, that's right.

So, I would encourage you to check out what your state has to offer as each state has different incentives and rebates. In general terms, you can bring the cost to your system down to about $14,000 when everything is all said and done.

Now, assuming that the average electric bill runs to about $1,200 a year (of course, this may be a lot more depending on the part of the country or the world), you can essentially save this entire amount of money per year if you draw your entire power sources from your solar-powered system.

In total, you can save roughly $30,000 over the lifespan of your system. That's not chump change if you ask me.

This last fact is particularly eye-opening when you consider that electric bills rise, on average, 5% per year. Moreover, this is a lot worse when you factor in that your electric bill is at the mercy of international oil prices and the earnings of the stockholders of the utility companies. Even if your local utility company is public, they still have to raise rates in order to pay their own bills.

Another very important reason for you to go solar is that it will increase the value of your home.

Think about that for a moment.

If you are shopping for a home, finding a home with a fully-functioning solar power system is a real plus. Purchasing a home that is energy independent is such an appealing factor that homebuyers are generally willing to bid against each other. This is especially true in markets where electricity is expensive. This is magnified when you consider how much electric bills can run up during the wintertime. When you factor in the cost of heating during the winter, having a solar power system will certainly alleviate much of the pressure on your utility bills.

In general, having a fully-functioning solar power system can add 3.6% to the value of a home, according to the National Bureau for Economic Research. Bear in mind that this is a *national* average. This is what I mean when I say that the price of your home can rise dramatically depending on the area in which you live.

Everything sounds really good at this point, doesn't it?

Yes, it is clear that we are extolling the benefits of having your own solar power system; there are a couple of caveats to keep in mind.

First, it is certainly worth taking your time to plan out how you will set up your system. While we will cover this in the following chapters, I would encourage you to walk around your home and visualize where you may choose to install the various components of the system: the panels, the battery, controllers, and so on.

The other caveat that I would like to point out is leasing.

There are some companies out there that offer the leasing of such systems. In short, leasing means that you pay the leasing company for the use of the system and not for the energy generated by the system. I am against this type of system for one *very important* reason: you will never be self-reliant. This is due to the fact that the system is not your property but the property of the leasing company.

Sure, it might be a lot more convenient to have the system installed by someone else, but when it comes time to rely on the system, the leasing company can pull your contract at a time. If you have ever leased a car, then you know what I mean.

Thus, I would not advise you to lease a system. It is always best to go through the wringer and set up your own system, which you know will always be there for you, no matter what.

That is why this book is intended for the do-it-yourselfer who is concerned about security, family's comfort, and safety. By going down this road, even if it is a bit tough, you will be able to get some peace of mind in knowing that your supply of power will always be there.

Unless the sun suddenly burns out next week, you can be confident that you will have a consistent supply of power. In the event of an emergency, this can be a real lifesaver. It doesn't matter if you use your system as a complement to your current electrical installation or to go completely off the grid. However, you choose to set up your system, you can be sure that you will never run out of power.

If you are still unsure about doing this, I would only encourage you to read through this book first. Once you have gone from cover to cover, you can then make up your mind if this is the right way to go for you.

Of course, it could be that going off the grid is not right for you.

And that's fair.

Chapter 1: Calculating the load

Quick electricity explanation for beginners

Electric current – a movement of electrons in a conductor of an electric circuit.

Conductor – is material where electrons can move easily and allow an electric current to move from one point of the conductor to another. (steel, copper, and so on).

Insulation – material that doesn't allow electrons to move easily (for example plastic).

To power things, we have to create pathways for electrons (electric current). This is called an **electric circuit.**

Measuring of electricity:

Electric Current (Amps) – the amount of electricity "flowing" in the wire. To conduct a lot of electricity (Amps), we will need bigger (thicker) wires, which we will determine by using the Amps rating. This metric could be present if electricity is flowing or is consumed by an appliance.

Electric Voltage (Potential) (Volts) – the size of the force that sends electrons through a wire. Energy potential is always present in an electric circuit.

In most cases, volts rating determines the compatibility of components. For example, if we have a 12-volt battery it can only power 12-volt appliances, and we should connect other components that have a 12-volt rating.

Electic Power (Watts) – unit of electrical power equal to 1 Amper under the pressure of 1 Volt.

Volts and Amps rating determines the Watt rating:

Watts = Volts x Amp

1W = 1V x 1A

A solar panel that produces 8.33 Amps of electricity at 12 Volts – producing 100 Watts

100W = 12V X 8.33A

A laptop that consumes 4.7 Amps of electricity at 19 Volts – consumes 89.3 Watts

4.7A x 19V = 89.3 Watts

Watts shows the total amount of electrical power a component of a system is consuming, storing, or producing at the moment.

In this first chapter, we are going to go through one of the most important parts of installing your solar power system: **calculating the load.**

Well, I would encourage you to start by taking an inventory of everything you wish to power by your system. If you plan to power your entire home, the size of the solar power system needs to be larger, you will need a much stronger charge controller, a solid inverter, and multiple batteries to store energy.

Let's take a look at each one of these components and how they come into play within the entire solar power system.

Solar panels

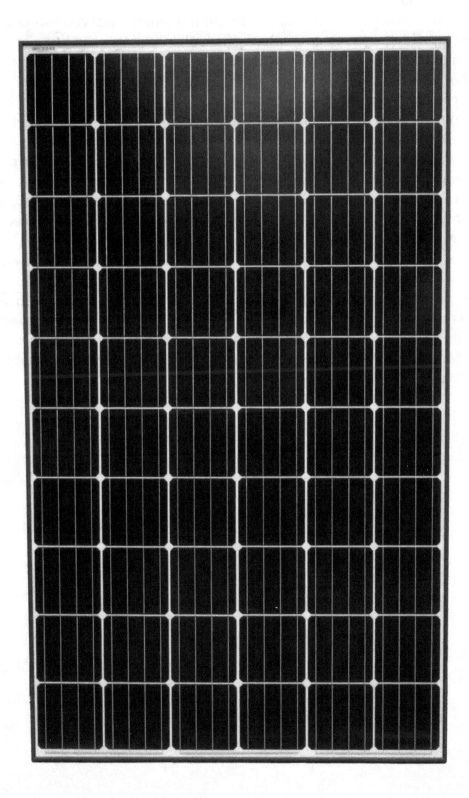

When you think of a solar power system, you might think that it just consists of a bunch of solar panels taped together. Once you have them all set up, you can just start plugging stuff in. In a nutshell, that's how it works, but it is not quite as simple as that.

Considering that the solar panels themselves come in all shapes and sizes, you can adapt them to the particular geometry of your home or vehicle.

A solar panel contains cells. Each solar panel cell produces about 0,5 volt. To get more than 0,5 volts cells are wired in series. The more cells a solar panel has a higher voltage of the panel. When multiple solar panels are wired together we get a solar array.

Now, it is not required to place the panels up on the roof of your home, though this is the smartest place to put them as they will be out of the way and will have exposure to the greatest amount of sunlight. That being said, it is possible to have the panels placed on other parts of your home, such as the backyard. The issue with placing panels in a backyard is that they will take up quite a bit of space. Consequently, that may not be an ideal solution.

Other potential configurations may include building additional structures, such as a raised deck, in order to expose the panel to sunlight while not having to sacrifice space. We will be taking a deeper look at these configurations in chapter 6. For now, I would encourage you to get an idea of what dimensions the solar panels may have and what the best place for their mounting would be.

Charge controller

Image source - victronenergy.com

In essence, the charge controller will regulate the amount of voltage generated by the panels. It is important to keep this in mind as the panels themselves may end up generating much more voltage than the usual 12 volts required for charging batteries.

So, the charge controller will make sure that the voltage levels are the required ones, thus ensuring optimal charging voltage at all times. In doing this, you will keep your batteries from getting fried. Also, you will be able to make sure your system is running optimally at all times and last long.

It certainly pays to look into what options you have at your disposal in terms of charge controllers. If you choose to run a much larger system, you will need to look into a charge controller that would be able to handle the load. This condition will play into the spot you will choose to install the charger itself while also factoring into costing considerations.

Inverter

Image source: SMA Solar Technology AG

Many folks ask me why an inverter is needed. The answer is actually fairly straightforward. Solar panels generate electrical power on a 12 volt DC current. However, most home appliances and devices run on 120 volts (or even 220/240 volts in Europe) and are made for AC current. So, the inverter is needed in order to ensure that such appliances get the power they need.

Now, if you are only looking to power a few lightbulbs or any other 12-volt appliances, then 12 volts system is more than enough. Therefore, there would be no need for an inverter. However, an electric range would normally run on 110-120 volts/ 220-240 volts. Consequently, the inverter would be needed in order to power the range.

Again, the ultimate decision boils down to how much you are looking to power in your home. So, it certainly pays to plan out all of these details before embarking upon the purchase of hardware.

Batteries

Image source: outbackadventures.net.au

The need for batteries comes down to the simple fact that sunlight is not available 24 hours a day. So, the batteries are the storage of energy when there no sunlight available such as during nighttime.

The batteries in this system essentially function like large car batteries. So, what actually happens during the use of your solar power system is that the panels charge the batteries, and then the use of the energy drains the batteries. As such, the batteries being constantly drained and recharged.

The battery only needs to store enough power to get the engine to crank when it is off. Once the engine cranks and starts running, the engine, through its gasoline power, runs the alternator, which then keeps the battery charged. This ensures that all of the electrical components of the car function properly.

Thus, a solar power system basically works the same way. If you have a solar power system without storage batteries, then what would end up happening is that the entire system would shut down once the sun goes down.

This is why determining your load will determine the size and number of batteries that you will need.

If you are looking to power an entire home, that is, a fridge, range, microwave, and lights, you will need several larger batteries. This will ensure that you will always have enough power to light up your entire home.

Predesigned mobile solar power systems

If you don't want to do calculations, and just want to install a solar power system quickly for your vehicle you can use these precalculated systems.

Backpacking system

Suits well, if you want to power with solar power a small vehicle, and can't install solar panels and battery bank in the vehicle or if you are traveling without your vehicle. You need to power mobile appliances like a phone, a camera, a small laptop, and other tools used while traveling.

We need:

- A fold-up solar panel with USB cable output (20 - 40 watts) to charge a battery;
- A few USB batteries with fast charge capability;
- USB AC adapter (also fast charge) to charge a battery with the grid when charging with a solar panel is impossible;
- To charge a laptop we can use a portable external battery charger, or buy a laptop that charges with USB;

Make sure that a battery and solar panel have the same voltage, and are suitable for charging by the solar panel. Read manufacturer instructions.

For car or minivan (200-watt installation):

If you plan to power small appliances like fan, laptop, USB charger, and so on;

Required components:

- 200-watt solar panels;
- 150 amp hours, 12-volt batteries;
- 20 amp MPPT charge controller;
- 750 or 1000 watt inverter;

For Van or RV (400-watt installation)

This system can handle a large laptop, TV, fridge, lights, fan, and so on. This setup is suitable for most people.

Required components:

- 1500 watt inverter;
- 400 watts of solar panels
- 40 amp MPPT charge controller
- 12 volts, 200 amp-hours lithium-ion battery which will weight much less than the lead-acid battery (300 to 350 amp- hours lead-acid battery – about 200 pounds (90 kg));

Bus or Large RV (800-watt installation)

For this system, we can use large appliances.

Required components:

- Inverter 2000 watt or more;
- 800 watts of solar panels;
- 12 volt, 300 amp-hour lithium-ion battery bank or larger 500 amp-hour lead-acid battery or more;
- 80 amp MPPT solar charge controller;

For homes and buildings, we have to calculate the solar power system manually.

Solar power system design method

1. **Estimating daily electricity consumption** (watt-hours consumed by your appliances in 1 day);
2. **Calculating battery bank size** (we can do this by knowing the estimated daily consumption);
3. **Calculating the number of solar panels** (for this we need to know the battery bank size);
4. **Calculating solar charge controller size** using solar panel array size;

Calculating the daily electricity consumption

Now, let's look at a simple means of calculating your load.

Step #1:

Make a list of all of the appliances and devices that you will power. This includes lights, water heating systems, and even air conditioning. List all of the items that you are looking to power in your home. If you are looking to power your entire home, then you might want to go room by room, making a list of all of the items that will be powered. I know that seems like a lot of work, but it should give you a fairly accurate assessment of how much power you will need.

Step #2:

Once you have come up with the list of all of the appliances, lights, and devices that you will include in your solar power system, it's time to **figure out how much power each item will require.** Most appliances, such as refrigerators, ranges, and microwaves, will have a sticker on them, which indicate how much energy they consume. Generally, their energy consumption is expressed in watts (kilowatts if an appliance consumes a lot of power – 1 kilowatt means 1000 watts) or amps and volts.

If you have given only amps and volts you can calculate watts simply by multiplying them.

Amps x Volts = Watts

So, to calculate the entire power consumption for the solar power system, you would need to make this assessment for each of the items you wish to power.

If you can't calculate how much an appliance consumes, you can visit the manufacturer's website to get that information.

Decide how long you plan to use each of them daily (hours).

When we know how much power consumes each appliance and how long we will use each of them. We can **calculate the total daily consumption.**

Appliance1 + Appliance2 + Appliance3 + ApplieanceN + … = total daily electrical power load (amount of electricity needed each day);

N – number of appliances;

Now, let's assume the following:

- One LED light consumes 5 watts, and you will need to power this light for an average of 6 hours at night. Thus, the calculation works out to 5 watts x 6 hours = 30 watt-hours total.
- One TV consumes 50 watts of power. You plan to watch TV for an average of 2 hours at night. That would work out to 50 watts x 2 hours = 100 watt-hours total.
- 1100 watt microwave used for 20 minutes a day will consume 367 watt-hours of power (1100 watts/ 3 = 367 watt-hours), 20 minutes is 1/3 of the hour.

- One refrigerator consumes 60 watts of power. Since the refrigerator will be powered on for 24 hours but actually will run for 4 hours because compressor activation is intermittent. It means that power consumption for the refrigerator will be:

60 watts x 4 hours = 240 watt-hours total.

Step #3:

Add them all up:
Light - 30 watt-hours;
TV - 100 watt-hours;
Refrigerator - 240 watt-hours;
Microwave – 367 watt-hours.
Total = 737 watt-hours.

This is the number of watt-hours that you would need to power all these appliances **for 1 day of normal use.**

Now we know that we need at minimum 737 watt-hours capacity battery bank.

If you happen to have more power-hungry appliances and devices, you might want to consider adding panels and battery bank or/and switching over to some less-consuming devices. Also, a great tip here is to switch over to LED lights as they usually deliver the same level of brightness without consuming nearly as much as a compact fluorescent bulb, not to mention an old-school incandescent bulb.

I have also seen other folks power their more power-hungry appliances such as a refrigerator, range, water heater, washer, and dryer on a solar power system. This alone would represent a considerable reduction in your power bill.

Other folks have the solar power system as an emergency backup generator-type system. This is intended to protect them from a potential power outage, especially in areas that tend to be hit by disasters such as hurricanes. Thus, the maximum load on this system isn't quite as robust, but rather, it is intended to power essential services such as refrigeration, cooking, lighting, and hot water systems.

In this diagram, you can observe the general schematic of how a solar power system is generally set up. You can tweak this configuration to suit specific needs. For example, you may do away with the DC load. Nevertheless, I wouldn't advise you to mess around too much with this setup. It has been proven to work.

One common question I get from folks pertains to the use of AC and DC.

Firstly, AC refers to the alternate current. It has been proven that AC is much more efficient over longer distances. As such, electricity generation plants output their energy in DC and convert power into AC and then subsequently transmit them over long distances via power lines.

This enables a reduction in power loss while ensuring that customers get as much power as they need for their homes. Most of the major appliances you use run on 120 volts (220/230 volts - Europe).

So, what you will find is that your solar power system will produce DC electricity, which then needs to be converted into AC so that your appliances will run as they have been built for AC and not DC.

There are a lot of devices that run on DC such as cellphone batteries that require DC power for charging. But this is something that you need not worry about as your phone's charger comes with an AC-DC converter built-in. Also, bear in mind that the batteries and solar panels you will use to produce and keep your electrical power – typically produce a 12/24/36 volt DC power. So, you would need the use of the inverter to switch it to 120 volts (220/230 volts) AC power.

What tools are required (basic tools)

1. You don't require to have all these tools. But to build any size system you will need to use most of these tools if not all of them.
2. Electrical tools: Wire strippers (automatic wire strippers up to 10 gauge wires, also there are strippers with adjustable blades – removes any insulation, if very difficult to remove insulation we can use razer blade tool), wire crimpers (3 point crimpers – up to 4/0 AWG cable (huge cables), hammer type crimper), cable cutters;
3. Standard tools: socket set to ratchet down nuts and bolts, a screwdriver, impact gun, drill bits, a marker to mark wires for cutting.
4. Hardware: screws, cable clamps(for holding a cable), various size washers, circuit breakers and fuses, a busbar, copper terminal connectors, soldering iron with solder, standard size crimp connectors and large gauge wires connectors, xt60 connectors.

Wires required

5. For the roof, you have to use a solar panel hook up wire (input wire) for the roof that has UV resistance and will last a long time.
6. For connecting 12 Volt DC appliances will be enough to use 10 or 12 gauge wires.
7. For connecting standard appliances will be enough to use 4 gauge appliances, 2 or 0 gauge for investors that have more than 2000 Watt or if the invertor is far from a battery.

Summary

In this chapter, we went over the most important step that you need to keep in mind before you even begin to look at the hardware: planning. If you plan your system effectively, you will drastically reduce the number of issues you will encounter when installing your system and then actually running it. So, I would greatly encourage you to take the time to plan your system carefully.

Chapter 2: Choosing a battery

In this chapter, we will be taking a look at one of the most important components of your off-grid solar power system: energy storage. Now, it should be said that batteries, on their own, are not indispensable to a solar power system. After all, the system can work perfectly fine without them.

However, the reason for having batteries is that sunlight is not available 24 hours a day, 7 days a week. As such, there is a clear need to have a device that can store energy during the night and cloudy days. This is why an off-the-grid system needs to have a solid energy storage system.

Without a solar power storage system – you use it or you lose it.

Batteries types:

Flooded Lead Acid Batteries:

Have liquid electrolyte or acid that needs to be monitored, and outputs hydrogen gas, which required good ventilation and checking water and gravity of the electrolyte. Advantages are high lifespan if maintenance is correct, low price, but need maintenance, and can cause a fire because of hydrogen gas if ventilation isn't proper.

Sealed Lead Acid Batteries:

Most common are AGM (Absorbed Glass Mat) batteries that have a silica glass mat electrolyte, and Gel batteries (silica gel electrolyte), this type of batteries are safer - non-spillable and don't produce gas, don't require much maintenance, but are more expensive and have a shorter lifespan.

Lithium-ion batteries:

Don't require much maintenance, have a high usable capacity, high lifespan, safe, has high cycle life but is the most expensive.

Now, the decision on batteries depends on the daily consumption that you have determined. As such, if you have calculated high consumption (watt-hours), then you will need multiple batteries that can hold the charge and slowly discharge as per the needs of your home or a vehicle.

Car batteries are not designed to discharge over long periods of time.

They have a low depth of discharge rating (5% of capacity). It means that 100 amp-hours battery can be safely discharged only to 95 amp-hours of capacity, and are designed to give a lot of current in a short amount of time to start an internal combustion engine.

Lithium-ion batteries such as the ones found on laptops or even electric vehicles are quite new technology compare to lead-acid that exist for 150 years now. Lithium-ion batteries able to keep large amounts of electricity, and they can discharge down to almost 0% if BMS (battery management system) allow this - without damage to the battery. To increase the life of the battery – discharge it to 50%. Unlike **a lead-acid battery, which never should go lower than 50% to ensure that the battery will last and not die in a year or two, to increase life you will need to discharge to 80% of battery capacity.**

Lithium-ion batteries are very expensive compare to lead-acid batteries. Price can be three or even four times higher than for the lead-acid batteries.

A battery bank for solar power systems should have *a high cycle life*, (number of discharges and recharges), long lifespan (up to 25 years), and big depth of discharge (how much electricity we can safely pull out of the battery);

For solar power systems the best suits deep cycle batteries that have all the required characteristics above (**AGM sealed deep cycle batteries and lithium-ion batteries (lithium iron phosphate)**).

Lithium battery lasts from 3 to 10 times more than an AGM sealed batteries. A lithium battery can be discharged from 5000 to 8000 times when a lead-acid battery only 500 to 800 times. But lithium batteries cost from 2 to 5 times more than lead-acid batteries.

If you do not plan to store and use a lot of power and don't want to spend a lot of money then AGM sealed lead-acid batteries will work well for you.

Lithium-ion deep cycle batteries are the best option to choose if money is not a problem for you and you need to rely on batteries on a day to day basis.

Estimating backup power (a battery bank size)

Backup power is needed for compensating for reduced power production during rainy or cloudy days and wintertime.

For mobile solar power systems, we need at least 2 days of backup power.

For stationary solar power systems that installing in cabins, tiny homes, or other buildings from 2 to 5 days of back up power is recommended because we can add any size and weight battery bank in comparison to vehicles.

Our calculated daily load is 800 watt-hours of AC power, which will power the appliances we mentioned above. For converting to AC power we will use an inverter that has efficiency let's say 95% or (0,95).

To calculate our daily power consumption that will take into account an inverter efficiency, we will have to:

Calculated daily load divide by the efficiency of the inverter (0,95);

800 watt-hours / 0.95 = 842 watt-hours;

Also, we can take into account the temperature where your battery bank will work. But it matters if you want to be very accurate and know where your batteries will be installed. There is a list of multipliers for each temperature, you can find them online. The ideal temperature for most batteries is about **25 or 26** ⍰.

Daily power consumption 842 watt-hours of AC power x 2 days of autonomy = 1684 watt-hours of backup power for mobile solar systems, and 842 watt-hours x 5 days = 4210 watt-hours for a stationary solar power system.

Of course, your daily consumption can be 10 or 20-kilowatt hours if you live in a hose then you will need a really big solar power storage.

Let's assume that we design the solar power system for a tiny house or a vehicle where you use only basic appliances.

We need a battery bank that has a useful capacity of 1682 watt-hours of power. When fully charged we can discharge batteries for 2 days without using solar panels to recharge them.

If we decide to use a lead-acid battery that we can discharge safely to 50% without damaging the battery.

We will need the battery that has at least 2 times bigger capacity (3368 watt-hours) than required to be able to discharge the battery to 50% and use 1684 watt-hours of battery capacity by appliances;

1684 / 0.5 = 3368 watt-hours;

To maximize the battery life span we will need to discharge the deep cycle battery bank only to 80%. And build 8420 watt-hours battery bank capacity.

1684 watt-hours / 0.2 = 8420 watt-hours battery bank size;

How to determine the Amp-hour rating of your battery bank?

Battery Bank Capacity (Watt-hours) / Voltage of Battery Bank System we Plan to Build;

8420 / 48volts = 175,41 Amp-hours;

If we decide to use a lithium-ion battery with a BMS (Battery Management System) that can be discharged safely. In this case, we can just build 2 000 watt-hours battery bank;

We can buy two 100 amp-hours, 12-volt batteries, and connect them in parallel (in this case voltage will be the same 12 volts, but amp-hours will add up to 200 amp-hours). The total capacity of the solar battery bank will be 2400 watt-hours (12 volts x 200 amp-hours = 2400 watt-hours) that are on 700 watt-hours (almost 1 more day capacity than we required) so our back up capacity is about 3 days without charging the batteries;

Battery bank sizing example (8420 watt-hours capacity (175,41 amp-hours), 48 Volts system):

1. After we determined the total amp-hour rating of the battery bank we have to decide on a **minimum battery capacity (individual amp-hour rating of a battery).**

For this, we have to divide the **total amp hour rating** of the battery bank by the **number of parallel strings we want to have** (the best for well-functioning battery bank we need not more than 3 series strings connected in parallel – recommended by experts).

Let's say we want to have **two parallel strings** for our battery bank. So, N = 2 strings;

Amp-hours / N = 175,41 Amp hours / 2 = 87,7 Amp hours;

We need a battery that has at least 87,7 amp hour rating.

By knowing this number we can choose for battery bank a 12 volt 100 amp hour battery.

2. **Determine the number of batteries in each series string.**

Battery bank voltage that we want (48 volts) / each battery voltage (12 volts);

48 volts / 12 volts = 4 batteries;

3. **Calculating a total number of batteries for the battery bank:**

Number of batteries in each string x Number of strings;

4 batteries x 2 stings = 8 batteries;

So we will have to buy 8 batteries (12 volts, 100 amp-hours), and create 2 series groups (4 batteries connected in series each), then connect them in parallel. How to do this we will discuss how to do wiring in chapter 7 (Battery bank wiring). The diagram below shows how it should look like.

Conecting 2 Series Groups in Paralel

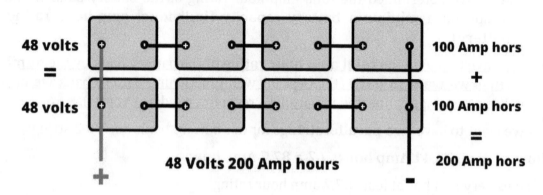

We have built a 9600 watt-hours battery bank. (48 volts x 200 amp hours = 9600 watt-hours)

9600 what-hours capacity is slightly bigger than we required in beginning (8420 watt-hours). So, we will be able to store enough energy for the daily use of appliances.

What to look for in a battery

While you could run your solar power system on larger lithium-ion batteries, these will end up costing you more and will not be the most accessible option available to you.

So, the best option that I have seen work for many folks is **AGM sealed lead-acid batteries**. Sealed batteries don't require much maintenance, they don't need to be refilled with distilled water, and we can hide them with an access panel.

Lead-acid batteries have two main advantages:

1.   **Cost**. AGM sealed lead-acid batteries are fairly large in size and have a much smaller capacity as compared to lithium-ion batteries. However, the biggest difference is the price. So, you can conceivably set up your system at a reduced cost right away by using lead-acid batteries.

2.   **Reliability**. They are a tried and tested way of storing energy. Many folks have them over decades, proving the effectiveness of lead-acid batteries. Therefore, you can rest assured that your solar power system will work appropriately by using lead-acid batteries.

Now, it is possible to have a combination of lithium-ion and lead-acid batteries, but not recommended. However, if you begin mixing and matching the types of batteries you use, you may not be able to ensure the full capacity of your system while reducing the overall efficiency of the system.

Capacity and power

In the first chapter, we determined the amount of power in terms of watts per hour. Of course, you may have noticed that your electric company bills you in terms of kilowatt-hours. The reason for this is that electrical energy is generated in vast amounts by power plants. So, residential customers are measured in kilowatt-hours based on the entire energy consumption of your home. As such, if you are looking to power your entire home, you can certainly take a look at your electric bill and see how many kilowatt-hours you are being billed for on a given month. You can always look at the average consumption of your home over the last six months. **That will give you a fairly decent assessment of the capacity that you will need for your batteries.**

Consequently, high electricity consumption will require installing multiple batteries to store enough power so that you could comfortably use all the appliances that you plan to use each day.

One tip to keep in mind: if you are powering your entire home, it is a good idea to have a breaker box in which you can just reset breakers or switches for your home's electrical system. Even if nothing is plugged into an outlet or if a light is off, it still consumes tiny amounts of energy, and you can turn off some parts of your home if needed.

When you go shopping for batteries, look for their storage capacity in terms of amp-hours. **Typically batteries are rated in amp-hours and volts.** The math here is simple. If your daily consumption is 5 kilowatt-hours, you will need to build from 5 to 25-kilowatt-hour (4 days back up power) battery bank if you choose lithium-ion batteries.

To calculate watt-hour capacity for a battery just multiply the amp-hour rating and voltage.

For example, you have found 12 volts, 300 amp-hour a lithium-ion battery bank.
The watt-hour capacity:

12 volts x 300 amp-hours = 3600 watt-hours, or 3600/1000 = 3,6 kilowatt-hours
To calculate the number of batteries that we need for 5 days of power consumption we will need to divide 25 kilowatt-hours by 3,6 kilowatt-hours = 7 batteries.

So, we will need to buy **7 batteries or 8 batteries** to create a battery bank for our solar power system.

There is another factor that affects the battery bank capacity and lifespan – the rate of charge and discharge. Typically a deep cycle battery requires (20 or 10 hours current) for optimal discharging or charging.

For example a 100 Amp-hour battery bank:

100 amp/10 hours of optimal discharge= 10 amps is our optimal charging/discharging current;

Typically, in specs for batteries is given the charging current for battery bank;

If you will charge or discharge batteries with higher current (amps) then you will reduce the battery lifespan and decrease the battery storage capacity. The lower the charging/discharging current the higher capacity the battery bank can have.

By knowing this you might want to adjust the battery bank size to provide good charging and discharging rates to have enough capacity and increase the lifespan of the batteries.

Depth of discharge

The depth of discharge is the fraction or percentage of the capacity (power) which has been removed from the fully charged battery. It shows a battery state of charge.

Let's assume that you consumed 5 kilowatt-hours from your battery bank. Since you have a full capacity battery bank of 25 kilowatt-hours, so **the depth of discharge can be calculated this way:**

5 kilowatt-hours / 25 kilowatt-hours = 0.2 x 100% = 20%

The depth of discharge is 20% for this battery bank.

Or, for example, if you have a 200 amp hour battery and it is discharged for 30 minutes at 100 amp current then the **depth of discharge can be calculated this way:**

100 amp x 30 minutes / 60 minutes / 200 amp hours = 0,25 x 100% =25%

The depth of discharge is 25%.

Round-trip efficiency

This point refers to the ratio of energy you put into the battery and the energy you get out of it.

No energy conversion is 100% efficient.

For instance, if you put in the battery 5 kilowatt-hours of power, but if you only get 4,4 kilowatt-hours out of it, **then you have an efficiency of 88% (4,4 kWh / 5 kWh =0,88 x 100% = 88%).**

Rtf = (Energy out of the battery / Energy you put in the battery) x 100%

In the battery, there are at least 2 conversions -electrical to chemical and chemical to electrical.

In the inverter when you convert DC to AC there are also some losses.

Battery life and warranty

Battery manufacturers will guarantee your batteries over a given period of time or a given period of cycles.

Let's start with cycles.

By "cycles", we mean the cycle of charging up a battery, then draining it. Each cycle is generally measured from a 100% charge down to about a 10% charge for lithium-ion batteries. It doesn't count the depth of discharge. Even if you discharged the battery to 90% and charged it back to 100% it will be a battery cycle.

As such, a manufacturer may rate your battery as good for 5,000 cycles. So, if you run through one cycle every day, you can consider that your batteries will last for about 5,000 days. That's actually not bad when you consider that 5,000 days is about 13 years.

However, you may find that the manufacturer may guarantee the battery for 5 or 10 years. The reason for this is that over time, the battery begins to degrade. **Also, you need to follow the battery's maintenance schedule, as indicated by the manufacturer.**

In this regard, your batteries will require the same type of maintenance as a car battery. You would need to ensure that the terminals are clean, the wires are not rusted or rotting, and that you are not draining them all the way down to 0.

So long as you stick to whatever the manufacturer indicated, you'll be good to go.

It should be noted that with lead-acid batteries, it is not generally necessary to change any of the internal components or replace the acid in the battery during its guaranteed lifetime. Nevertheless, there are cases in which there is internal damage to the battery's physical cells, which will render it inoperative.

One other type of warranty on batteries may come in the way of the charge they hold. For instance, a battery may be under warranty for five years and may be replaced by the manufacturer if its maximum charge falls below 70%. In this case, the manufacturer may replace it under warranty or perhaps rebuild it.

For instance, a battery might be warrantied for 5000 cycles or 12 years at 70% of its original energy capacity. It means that in 12 years, your battery will lose not more than 30% of the original ability to store the power.

To maximize a lead-acid battery life we will have to discharge the battery bank only to 80% of its capacity. Depth of discharge is 20%. In this case, we will maximize the number of charging cycles **up to 3200 cycles instead of 800 cycles if you would discharge the batteries up to 50% (numbers change depending on the batteries you have).**

Personally, I would go for the longest-rated battery I can find for deep cycle sealed lead acid battery or deep cycle li-ion batteries. So, if I can find batteries that are guaranteed for 10+ years, then I would go for this choice.

Manufacturer

Last but not least, it is vital to consider the manufacturer of your batteries. I will not get into details about the actual manufacturers themselves, as I want to avoid it seem like I am making a plug for anyone.

Regarding this specific point, I would advise against going for the cheapest option. Because you may end up with batteries that don't last as long as they are guaranteed for even if you take care of them.

To choose a battery go on Amazon and search for deep cycle batteries for solar power systems (Li-ion battery or AGM Sealed batteries). And look for reviews of people who already bought a battery. Typically the higher rating a battery has - the better quality a battery is.

Always check the manufacture date on the batteries.

Why?

Well, the longer that batteries are sitting on a shelf before use, the quicker they begin to degrade.

Summary

Choosing your batteries is a big decision. This decision will make or break your entire solar power system. So, I would advise you to do some shopping and compare. If you can talk to other folks who have been down the same road, you can get a sense of what works and what doesn't.

Requirements:
- Low maintenance cost;
- Long lifespan;
- Low self-discharge;
- High round-trip efficiency;
- High depth of discharge;

By learning from others' experiences, you can be sure to avoid making mistakes. Furthermore, try your best to invest in quality.

A battery bank for solar power systems should have a high cycle life, (number of discharges and recharges), long lifespan (up to 25 years), and big depth of discharge (how much electricity we can safely pull out of the battery);

For solar power systems the best suits deep cycle batteries that have all the required characteristics above (AGM sealed deep cycle batteries and lithium batteries (lithium iron phosphate)).

Lithium battery lasts from 3 to 10 times more than an AGM sealed batteries. A lithium battery can be discharged from 5000 to 8000 times when a lead-acid battery only 500 to 800 times by discharging them to 50% of capacity. But lithium batteries might cost from 2 to 5 times more than lead-acid batteries.

The choice is yours!

Chapter 3: Choosing solar panels

Just like choosing the right batteries, the decision regarding which solar panels to choose is equally important.

As such, we will take a deeper look at which solar panel you need to choose so that your solar power system can be effective and deliver the amount of energy that you need.

Now, the decision of which solar panels to choose depends on a myriad of factors. For instance, you would have to consider where they will be installed, what dimensions you can reasonably fit into the space you have designated, and the actual load itself.

Before we get into the specifics of the solar panels themselves, let's consider a few things.

Solar panel ratings

Solar panels have a series of specifications which you need to become familiar with:

- Peak Power (Pmax)
- Voltage @ Pmax (Vmpp)
- Current @ Pmax (Impp)
- Open Circuit Voltage (Voc)
- Maximum Series Fuse

20 volts solar module specifications (Example)

Maximum power (Pmax)	230W
Maximum power voltage (Vpm)	29.49 V
Maximum power current (Ipm)	7.80 A
Open circuit voltage (Voc)	37.20 V
Short circuit current (Isc)	8.39 A
Module efficiency (ηm)	14.3%
No. & type solar cells	60 in series/ 6"(156x156 mm) multicry
Maximum system voltage	TUV:DC 1000 V/UL:DC 600 V
Series fuse rating	15 A
Performance tolerance	±3%
Operating temperature	-40 to +90 ℃
Storage temperature	-40 to +90 ℃
Dimensions	1626 x 990 x 50 mm / 64 x 39 x 1.96 in
Weight	20.0 kg/44.09 lbs
Output Terminal(Tyco J-Box)	1394462-4(-)/6-1394461-2(+)

Image source: enerzytech.com

These specifications are listed under Standard Test Conditions (STC).

The panel was tested under these conditions: radiation (Amount of solar light energy falling on the given surface - of 1000 watts per square meter) - 1000 watt/m², air mass - 1.5, and cell temperature 25° C.

What this means is that if you install these panels under real-world conditions, for example, where temperatures exceed 40°C, then your panels will produce less power than under STC. The lower temperature of solar panels the better they will produce power. The higher temperature of solar panels the worse they produce power.

Now, let's go over each one of the components in the specifications.

- **Open Circuit Voltage (Voc)**. This voltage that a solar panel produces with no load on it. For example, if you would like to read a solar panel Voc you will need to use voltmeter and measure volts between positive and negative leads of the solar panel.

Using open-circuit voltage (Voc) we will determine how many solar panels we can wire in series to connect to the solar charge controller or inverter.

We have to remember that fuses and breakers protect from overcurrent and don't protect from overvoltage.

So we have to be sure that the total voltage of our solar array will not be bigger than the voltage that a charge controller or inverter (when you connect the solar array directly to the inverter) can handle without damaging them.

- **Short Circuit Current (Isc)** Amount of amps that a panel is producing when not connected to a load and under sunlight but when both positive and negative wires of the panels are connected to each other (short circuit), or circuit under conditions with no resistance.
- **Maximum Power Voltage (Vmp or Vpm).** In this spec, we are referring to the maximum voltage that the panel can produce when it is at its maximum power. So, let assume a Vmp rating of 36.7 volts. This means that when the panel is producing 255 watts, the panel's voltage will surge to 36.7 volts.
- **Maximum Power Current (Imp or Ipm).** The current that a panel produces when the power output is greatest. The panel that has a Pmax rating of 255 watts will produce a maximum current - Imp of 6.95 Amps.

- **Maximum Power (Pmax).** Peak Power, or Pmax, refers to the output for the panel under STC. **Pmax = Imp x Vmp;**

- **Nominal Voltage**
It is a category to know what equipment can be used together. It is not a real voltage that you can measure.

Nominal voltage for solar panels:
12V(36 cells), 20V(60 cells), 24V(72 cells)

For example, you can use 12-volt solar panels (that produce Vmp - 18 volts, and Voc about 22 volts) with a 12-volt solar charge controller to charge 12 volts battery that has an actual voltage of about 14 volts.
Each cell produces about 0,5 maximum power voltage (Vmp).

- **Maximum Series Fuse.** The highest current a solar panel can produce at no voltage. This refers to the biggest size fuse to use with a solar panel.

Position of the panels

Most folks choose to mount their panel on their roof. The main reason for this is that it provides the panels maximum exposure to sunlight. However, you need to consider the weight that your roof would have to bear, especially in the wintertime, when snow might weigh down on your roof.

So, the position of your panels may affect the overall amount of sunlight they get. This will definitely reduce the amount of energy they generate, and the amount of charge sent to the batteries.

Ideally, for stationary roof-mounted solar power systems, we should mount the solar panels at an angle that is equal to the latitude of the location where the system is installed.

More in detail we will discuss this in Chapter 6.

Irregular voltage

Regular power plants produce massive amounts of electricity, which is regulated before being sent to customers through transmission lines. As such, power companies try their hardest to keep a steady voltage. The main reason for this is efficiency. That way, no power is wasted, and electric companies can sustain their profitability.

In the case of a solar power system, the voltage is highly irregular. This is due to the fact that sunlight is not stable. For example, on a cloudy day, you might be getting a certain amount of limited sunlight. Then, suddenly, the clouds break, and the sun begins to shine brightly. This will provide a greater burst of sunlight and thereby electrical charge. This is where a charge controller plays a key role in avoiding surges of energy (volts, amperes) coming from solar panels and from overcharging the batteries and damaging them.

Types of solar panels

Essentially, there are three types of solar panels: monocrystalline, polycrystalline, and thin-film or flexible. The highest efficiency has monocrystalline panels and lowest efficiency thin-film panels.

Solar panels, depending on their size, will have a given output. As such, solar panels come in all shapes and sizes. You can have some which are a couple of inches wide and are used for charging cell phones and other small electronic devices, and you can have larger panels, which might be 65 (165.1 cm) by 39 inches (99 cm) with some variations, or 5.4 by 3.25 feet.

So, the main difference between monocrystalline and polycrystalline panels is that monocrystalline panels are a little bit more efficient and thereby more expensive.

What does it mean that they are "more efficient"?

In this case, "efficiency" refers to the rate at which the panels are able to convert sunlight.

Polycrystalline perform better with low light, require less space (are smaller), are black in color, efficiency – 17 to 20%.

But, if the cost is a consideration for you, the polycrystalline panels (are blue in color) would offer you a more cost-effective option though they do tend to be a bit less efficient (3% - 4%), and work worst with low light.

At the end of the day, if you live in a very sunny part of the world, you could acquire the cheaper panels as you would have access to more sunlight. However, if you live in an area that gets less sunlight, then you might want to spring for the more efficient panels (monocrystalline).

Thin-film panels have the lowest efficiency but can be used on the roofs of cars, RV's where aerodynamic matters for lowering fuel consumption of the vehicle and for lowering air resistance while driving.

Criteria for choosing solar panels

There are three, main criteria that you should take into account when choosing solar panels:

1. Efficiency
2. Durability
3. Quality

First of all, efficiency refers to the output of each panel. This is important based on the load you are looking to set up for your system. So, you can choose to go with monocrystalline panels for efficiency. This will deliver the most amount of energy as per the output printed on the label.

For instance, you have a 100-watt panel. It is safe to assume that your panel will not deliver the full 100 watts. Granted, this depends on a host of factors. Typically a 100-watt panel will have output from 70 to 100+ watts output depending on temperature, amount of sunlight, and type of solar panel.

It is also important for you to see what the manufacturer's warranty is on the panels' production. Most manufacturers will guarantee a give output threshold over the life of the panels' warranty. As such, the manufacturer may guarantee that your panels' output will not fall below 80% of the original output over the warranty period.

So, you might consider purchasing a few large panels that will deliver the output that you need, or a series of smaller panels that, when added up, deliver the production of electricity you need. Again, this decision boils down to the physical space you have to mount the panels themselves.

As far as durability is concerned, manufacturers will guarantee the lifespan of panels anywhere from 5 to 25 years. Naturally, the cheaper panels will have a shorter lifespan, while the more expensive ones will have a longer lifespan.

In general terms, panels are pretty durable. Unless they are subject to extremely harsh weather conditions, they should hold up pretty well over time.

If cost is a consideration, you can go for the polycrystalline panels rated for 20 years. I feel that 20 years is a good round number. This can help you see returns on the investment you are making in the system. I believe that 20 years is a fair amount of time between upgrades and replacements.

Reputable brands and manufacturers will provide you with the best warranties. So, you can rest assured that quality panels and good batteries will keep your system running strong, especially if you choose to go entirely off the grid.

Solar array size calculating

For vehicles:

We can just fill the roof with as many solar panels as possible – usually, this will be the right thing to do because the roof space is limited, and to install a lithium-ion battery bank system because it usually can handle large charge rates, much lighter and requires less space on a vehicle than a lead-acid battery.

For houses and homes: typically we can install any size solar array and any type of battery bank.

- The lead-acid battery has a lower charge rate than a lithium-ion battery and wants to be fully charged every day. The solar array should be charging to the full in 3-5 hours of full sunshine.
- The charge rate of lead-acid batteries can be much higher if we connect multiple batteries in parallel instead of using 1 large lead-acid battery (exceptions for batteries designed for high charge rate).
- **Lithium-ion batteries only need to be fully charged every few months** (look in the battery manual);

A lot of _solar application batteries give recommendations for maximum and minimum solar array size and charge rate._ (look in a manual or call the manufacturer to know for sure).

Manual calculation of solar array size

Maximum size:

Find in a manual or datasheet (online) and **find a "maximum safe charging rate" in amps**. The maximum electric current produced by the solar array should be always less than the maximum charge rate of batteries.

Maximum Solar Array Current (amps) < Maximum Charge Rate of a Battery Bank (amps)

We already know that our battery bank has a capacity of 2400 watt-hours of power. (two 100 amp-hours, 12-volt batteries).
1 battery typically has about 40 amps maximum charge rate.

<u>The required maximum power that needs to be produced by the solar array can be calculated this way:</u>

Pm = I x V;

Where:
Pm – maximum required solar array power (watts)
I – maximum battery charge rate (amps);
V – batteries voltage (volts)

Pm = 40 amps x 2 batteries x 12 volts = 960 watts;

The 2 batteries' wired in parallel - the maximum charge rate is 80 amps (40 amps + 40 amps).

The number of solar panels (N):

Nmax = Pm/Pb

<u>Where</u>:

Pb – power generated by 1 Solar Panel, watts;
Pm - maximum required solar array power, watts;

If we chose 100-watt panels;

Nmax = 960 watt / 100watt = 9,6 panels

Let's round it to 9 panels. Because if we round it to 10 there is a risk that the solar array will produce more than the maximum charge rate (80 amps) for charging the battery bank.

If there is not enough space on your roof for panels, you can try to reduce the number of panels by choosing 200, 300, or ever 400-watt panels.

Make sure that the nominal voltage of the solar panel array is a little higher than the battery bank voltage or at least the same (12, 24, 48 volts). You can increase volts by connecting elements in series.

Minimum size solar array calculation:

We need to calculate the amount of solar power required to charge the battery in the worst-case scenario when you have very few Peak Sun Hours (in the wintertime). Typically it is about 3 hours. You can find this information online for your location.
We have to divide the daily watt-hours usage by 3 hours.

About 0,7 – is the efficiency of an MPPT solar panel system (wire loses, fuses, controllers, batteries losses, etc.)

842 watt-hours / 3 hours / 0,7 = 400,95 watts;

So we will need 400 watts solar panel array.

. Four 100 watt solar panels, or two 200-watt panels.

Solar panel buying recommendations

- Find panels that have MC-4 connectors.
- Buy solar panels in pairs (2/4/6/8/10) this makes wiring solar panels easier.
- For vehicles, the best option is 100-watt panels because they are strong, easy to install, easy to find, and relatively cheap.
- If you want your system to be as efficient as possible then buy monocrystalline batteries. They are more expensive than polycrystalline batteries but have a longer lifespan, higher efficiency, handle higher temperatures.
- If you want to minimize the weight of solar panels and maximize the aerodynamic of the vehicle, or if you have a curved roof and are not able to mount glass panels

you can find flexible solar panels they are much lighter than flat panels and can be bend.
- Flexible panels have smaller efficiency and a reduced lifespan compared to glass batteries.

Flexible panels are a little more expensive.

From the practice of using flexible panels, people noticed that often they will burn out in a few months or a year. Buy them with a warranty.

They can overheat and possibly cause a fire.

Most flexible panels can't bend more than 30 degrees.
- Most manufacturers of flexible panels say that you can flush them on the roof. But it is always good to have some space (ideally at least 10 cm) between a solar panel and a roof so that airflow is going between them and cool the panel.

Installation

Most folks would much rather have a contractor come in and install the panels for them. This is certainly valid, especially when you are not sure about how to do it.

While it is certainly valid to go this route, be aware that it will set you back a few extra bucks. A reputable contractor will charge you anywhere from a few hundred dollars to a few thousand dollars depending on the size and complexity of your system.

However, I must say that this is something that you can perfectly do yourself. We will be looking at the actual installation process in later chapters. At this point, though, I would like to advise you to make a decision on where you plan to install the panels.

Summary

If you have a limited budget, you can go for the more cost-effective options. This can certainly help you get your solar power system off the ground.

So, you might consider installing a smaller panel with one battery and starting off there. This will enable you to power some small appliances in your hose or vehicle (lights, laptop, fan). And when you decide to power bigger appliances such as a fridge or a microwave then you can build up your system, by installing more panels and adding more batteries. Either way, an incremental approach is a way that can help you get your system off the ground without having to make a sizeable upfront investment. But you will need to plan ahead by buying a bigger charge controller and inverter to be able to add panels and batteries without buying new equipment.

After all, you have nothing to lose.

In fact, you may see some real savings in your electric bill right away by simply using solar power for a fraction of your needs. Those savings can then be reinvested in building up the rest of the system in a such way that you can power most of your appliances and then achieve your goal of going off the grid.

Becoming energy independent is certainly a goal to strive for.

Chapter 4: Choosing the charge controller

In this chapter, we will be taking a look at an often-overlooked component of a solar power system: the charge controller.

Because of the changing amount of sunshine that is falling on panels, there will be spikes in the output voltage from the panels. Naturally, this is something that we want to avoid. And a solar charge controller should provide a constant charging voltage for our storage system (battery bank).

Having a charge controller is essential in order to prevent surges in voltage, protecting from damage to your batteries, increasing lifespan, and ensuring efficiency throughout the entire solar power system life.

Solar charge controller reduces the voltage of solar array to the charging voltage of the batteries that is about 14 volts. Most solar panels produce about 20 volts that is much higher than required for charging batteries. In the event of spikes in voltage, the charge controller (for example MPPT controller) will convert that extra voltage into amperes, thereby limiting the amount of voltage that the batteries will get without losing power. As we already know volts times amps equals watts. When the solar charge controller reduces the voltage from 18 volts to 14 volts it increases amperage and power on the output of the charge controller will be equal to power into the solar charge controller.

Furthermore, charge controllers may come equipped with safety features such as shutting off charging when voltage spikes up too high, or when batteries have reached max load.

A lot of charge controllers have over paneling protection that is also useful if you decided to add extra panels to your system to maximize power production.

Types of charge controllers

Now that we have established the need and importance of installing a charge controller in your solar power system, the time has come to look at the various types of charge controllers.

- Shunt - simple 1 or 2 stage controls (rarely used);
- PWM;
- MPPT;

The features of solar charge controllers:

- Monitors the reverse current flow that can lead to discharging batteries by solar panels - at night voltage of batteries (about 14 volts) will be higher than solar panels produces because the sun isn't shining;

- Protects the battery from overcharging – when batteries get fully charged the solar charge controller stops charging batteries that can lead to damaging batteries, and keep charging batteries to the full automatically;

- Reduce system maintenance;

- Provides over paneling protection

Also, a solar charge controller can have:

- Display and/or remote monitoring;

- Low voltage disconnect (LVD) – if the battery voltage gets too low, the charge controller will disconnect the DC load and prevents the batteries from discharging more than required or get damaged, the DC load connects to the charge controller;

- Temperature compensation to improve and optimize batteries charging depending on the temperature of batteries;

When you go shopping for your hardware, you will find these three main types of charge controllers Therefore, you need to become familiar with the differences between them.

- **Simple 1 or 2 stage controls (Shunt controller)**

This type of charge controller simply regulates the charge to the desired voltage. This solar charge controller has a shunt transistor that controls the voltage in 1 or 2 steps, simply shorts or disconnects the solar panels when a certain voltage appears.

It is the least efficient type of charge controller but very reliable, there are very few things that can break.

This type of solar charge controller is rarely used nowadays, so I will not speak a lot about them. But you still can see them on old solar power systems, or see some cheap controllers available online.

- **PWM**. (Pulse Width Modulations), efficiency is about 70%, which is another type of charge controllers. This type of controller essentially is about 20 to 30% less efficient than the MPPT controller which means that when we will use a cheap PWM controller on every 10 panels we will have to add another 3 panels to get the power that we would get by using an MPPT controller (additional few hundreds of dollars). This type of controller is affordable and a good solution for small solar systems - solar panels of which are producing power under high temperatures – between 45 and 75 degrees of centigrade. The controller doesn't work efficiently if solar panels are shaded. A 20 amp controller costs about 20 to 30$ but is less efficient than an MPPT controller.

To use this charge controller - the nominal voltage of solar panels and battery bank should be the same. A 12 volts panel should charge the 12-volt battery.

This charge controller simply reduces the voltage output of solar panels to the voltage of the battery bank without changing amps, and because of that, there are some power losses and lower performance.

- **MPPT**. The Maximum Power Point Tracking, or MPPT is the most efficient solar charge controller. It tracks the output of the panels in such a way that it can regulate the voltage and amperage and provide the best charging capability without losing power on the output of the solar charge controller. On average, this type of controller can deliver between 10% to 30% higher performance as compared to the other controllers. Efficiency is 95% to 98%. Furthermore, the MPPT controller can compensate for low irradiance levels (less sunlight) or cap the system when there is a spike in voltage due to high irradiance levels. As such, it is the most sophisticated type of controller and, therefore, the costliest. Nevertheless, it is certainly worth the investment and the type of controller you can set and forget about.

Can be used when the nominal voltage of panels is higher than the battery bank nominal voltage. For example, 60 volts (20 volts x 3 = 60 volts) solar array can be charging 48 volts battery bank. Or you can use this charge controller to connect a 20 volts solar panel to 12 volts battery bank.

The ultimate decision on which type of controller boils down to the following criteria:

- Solar array size;
- Cost;
- Efficiency;

If you are looking for the most cost-effective option, you can choose to use a simple 1 or 2 stages, solar charge controller. You need to keep in mind that it is the least efficient controller and almost no one uses it these days, so I don't recommend using them.

In the case of efficiency, the best choice is the MPPT. These outperform the other two types of controller's hands down. As stated earlier, it can increase efficiency anywhere from 10% to 30% which means that the time needed to charge the batteries will be less. This is ideal for climates where sunlight isn't as abundant or for larger systems, such as those powering an entire home, or if you have limited space for mounting solar panels.

As far as sophistication, you would have to decide how much technology you need in your system. A good, middle-of-the-road approach would be to install a PWM controller. This is a solid controller for mid-sized systems or in climates where there is abundant sunlight. PWM controllers also offer the best efficiency-cost ratio. While MPPT controllers are the most efficient, they are also the most expensive. So, it's up to you to determine whether your budget allows for it.

Personally, I would spring for an MPPT controller if I could afford it. Unless you are keen on keeping a close eye on your system on a daily basis, an MPPT controller will allow you to setup the system and leave it. I believe this peace of mind is well worth the cost. But, not everyone has the budget to go that way.

One other consideration, if you are incrementally building your system. You can spring for a PWM controller while your solar power system is smaller and then upgrade to an MPPT as your system grows. You might not think that it would make sense to invest in two controllers. But it is better to look at it in terms of upgrading your system as opposed to spending on the same item twice.

Rating of the charge controller

When shopping for the charge controller itself, it is important to become familiar with the rating on the controller, that is, the technical specifications on it.

It is worth noting that all three types of controllers have the same characteristic though their individual characteristics will vary according to their type.

The determining factor you need on your solar charge controller will be the voltage of the batteries you are using and the maximum output current of the solar array.

So, if you are using 12-volt batteries, then your charge controller should be outfitted for the 12-volt batteries.

If you are using 24-volt batteries, but your controller is set for 12 volts, then you will not be able to power the batteries.

On the contrary, if your charge controller is set for 24 volts, but you have 12-volt batteries, you may end up delivering more voltage than specified for the batteries. This will lead to damage. So make sure that your charge controller matches the voltage on your batteries and also the voltage of a solar array you build.

That being said, you can buy an MPPT charge controller, which can regulate multiple voltage levels. That would certainly be of benefit to you, especially if you upgrade from lower-voltage batteries to higher-voltage ones or decide to rebuild your storage system.

The cold weather (low temperature) increases the voltage output of solar panels and you must count this when choosing a solar charge controller. The voltage may increase up to 25% if in the place where you live temperature falls up to -30 degrees of celsius. So you have to make sure that your solar charge controller will handle this voltage input. It is always better to choose a bigger size solar charge controller for voltage increase and just in case you will decide to add more solar panels to your system.

If your solar charge controller should handle 100 volts then you should choose at least 125 volts charge controller (25% more).

The other thing to consider is the amp rating of a solar charge controller. A good way to calculate the Amp rating of a solar charge controller that we will need as follows: The Amp rating of a charge controller (I) equal solar panels array wattage divided by the voltage of the battery bank.

I =P/V, Amps

I – amp rating of the charge controller (amps);

P – total solar panels array wattage (watts);

V – voltage of the battery bank (volts);

For example:
The solar panel array power output is 400 Watts. The charge controller is reducing voltage to 14 volts, for charging the batteries. So, 400 / 14 = 28.57 Amps. This means your charge controller should be equipped to handle a load of 30 Amperes or more.

30A x 1,25 = 37,5A (25% for low temperatures increase - to be safe)

So we should choose a charge controller that can handle 37,5 Amps. So we can choose and use safely at least a 40 amps charge controller.

If your charge controller is not equipped to handle a current such as this, then you run the risk of damaging the charge controller.

When you will be buying your solar charge controller to **check the maximum input voltage of the controller** if your solar panels are wired in series and have high voltage. Typically maximum input voltage of a solar charge controller is between 50 to 150 volts.

I always make sure that the input voltage for the solar charge controller is at least 25% higher because I live in a place where the temperature can fall to -20 degrees of celsius.

A 100-watt panel in winter with low temperature can produce up to 120 watts of power when the sun is shining, and I don't want to damage the solar charger.

If in a place where you live the temperature never falls below 0 then you can not expect such an increase in voltage or amps for a solar array.

Choosing a solar charge controller

MPPT (maximum power point tracking) charge controller - can find the highest efficiency battery charge rate with solar array power provided, has efficiency from 95% to 98%, but are a lot more expensive than PWM controllers. 20 amp controller costs around 120$.
PWM (pulse width modulations) – efficiency is around 70%.
Costs around 20 to 30$.

MPPT controller can be about 20% more efficient than a PWM controller with the same number of panels. It means that when we will use a cheap PWM controller we will need to add 1 or 2 (additional 100-200 dollars) solar panels to get the power that we would get using the MPPT controller without buying additional panels.

If you have limited space on the roof for solar panels MPPT charge controller will be the best option for you.

I recommend buying a charge controller to which we can add a charge controller monitor to track the data and troubleshooting the solar panels.

You can add to your charge controller:
Temperature compensation – to determine the temperature of the battery bank, this will increase the life of the battery by allowing a charge controller to charge at a safe charge rate with the current temperature.
Charge Controller Monitor – shows how much power the solar array is producing (amps, volts, and more).

Decision making

Now that we have presented the benefits and drawbacks of each type of controller, the ultimate decision of which controller to use depends on how big your solar power system is meant to be.

If you are looking to power your entire home, and want to build the most efficient system I would urge you to spring for an MPPT controller. These types of controllers are the best option. During day time, they can be very efficient.

Nevertheless, if you want to build the system fast, and efficiency is not so much important for you or perhaps you are on the budget then you might be more comfortable with a PWM charge controller.

So, to find the solar charge controllers you need, go online, and compare different types and models with the best reviews and prices. I consciously don't go into specific models - I don't promote any companies here, because they might come and go.

Summary

The decision of which charge controller to use depends upon the voltage of the batteries you are planning to use, and the current flowing from the solar panel array, indicated by the amps.

As I have stated, I would personally prefer having an MPPT controller installed from the get-go. Although this may not always be the most cost-effective decision you can make when building your system.

Quick tip - First, we have to connect a solar charge controller to a battery, because if we connect it to the solar panels first we can damage the charge controller. And the opposite for disconnecting – first solar panels, second batteries;

Choosing a solar charge controller

Depending on the solar power array size we will determine the amp rating of the charge controller.

The Amp rating of charge controllers shows how much current it can produce at a battery bank voltage.

For example, a 60 amps charge controller can deliver a maximum charge in 60 amps.

The Amp rating of a charge controller (I) equal solar panels array wattage divided by the voltage of the battery bank.

$I = P/V$

I – amp rating of the charge controller (amps);

P – total solar panels array wattage (watts);

V – voltage of the battery bank;

For example:

If your solar array size is 800 watts and your battery bank voltage has 12 volts (charging voltage about 14 volts);

$I = 800$ watts$/14$ volts $= 57{,}14$ amps (minimum amp rating of solar charge controller);

Amps ratings of controllers - 10/20/40/60/80/100 amps.

To handle 57 amps of current we chose 80 amps a solar charge controller;

If you can afford to buy a larger controller that is needed and/or plan to add more panels to the system then buy a larger controller.

Also keep in mind, when you will be buying your solar charge controller to **check the maximum input voltage of the controller** if your solar panels are wired in series and have high voltage. Typically maximum input voltage of a solar charge controller is between 50 to 150 volts.

Chapter 5: Choosing an inverter

As stated earlier, the reason for installing an inverter lies in the way that the solar panels convert sunlight into electricity. The solar panel generates direct current or DC, but we need about 120 volts 60 Hz (or 230 volts, 50 Hz frequency if you live in Europe) AC electricity for powering appliances.

When electrical power generation became commercial at the beginning of the 20th century, power plants would produce vast amounts of DC power but quickly discovered that it couldn't be transmitted efficiently over long distances. This limited the coverage area that power plants had. In addition, the more power that plants were able to generate, the more that got lost in transmission.

Initially, this made electrical power unviable until Nikola Tesla discovered Alternate Current (AC). Eventually, Tesla's idea became commercially viable, and electrical power consumption spread widely around the globe.

Since then, virtually all appliances are powered by AC, as this is the type of energy that is used for residential customers. Moreover, smaller electronic devices such as cell phones and laptops run on DC. As such, cell phone and laptop chargers have a built-in converter that converts AC into DC so that these devices can run. Other types of batteries, such as the run of the mill Duracell and Energizer kind, also produce DC. The chemical reaction within the battery generates a given amount of voltage and current, which powers devices.

So, after a brief physics class and some history, you can see why the inverter is necessary.

One note: the system configuration which we presented way back in the first chapter had a DC output section. In this case, you could use this DC output to power devices that run on DC. So, it is certainly worth having this option available.

Types of solar power inverters

It is important to use an inverter that has been designed for solar power systems. It should be noted that the inverter receives its feed from the batteries. So, the panels charge the batteries through the charge controller, the batteries supply an inverter, and then the inverter converts DC electricity to AC. The current that goes from the inverter is what ends up reaching the outlets in your home or vehicle. After this, you will be able to power your appliances.

The component of your solar power system that will most likely to fail in the first 10 to 15 years is the solar power inverter. So, I would recommend choosing a solar power inverter that has the longest possible warranty even if it will be a little more expensive.

Off-grid inverters are inverters that aren't connected to the grid and have battery storage.

On-grid or grid-tied inverters can be directly connected to the grid and solar panels and don't require a storage system.

For domestic consumers, there are inverters from 800 to 10 000 watts.

Inverters also can be classified by voltage input the inverter accepts (12 volts, 24 volts, 48 volts, 96 volts) **DC power**.

In mobile solar power applications such as for vehicles most commonly used are 12 volts DC inverters. For residential photovoltaic systems, the most commonly used are 48 volts inverters.

Advantages of on-grid or grid-tie inverters

1. You can feed to the grid when the solar panels produce an extra amount of power and save extra money.

2. You can charge the batteries from the grid (the grid charger for the battery bank can be inbuilt in the inverter). When your batteries became drained and your solar panels can't charge it then you can charge batteries by using power from the grid;

3. Provides smooth power to the load (hybrid inverters). An inverter can get power from either the battery bank or the grid when solar panels can't deliver the load needed;

Thus, there are a few types of inverters:

*   **String inverters – grid-tied;**
    Multiple strings of solar panels are connecting to one string inverter and then the inverter converts DC power to AC power; the most commonly used type of inverters. The best choice if strings of solar panels don't get any shading then the system can work efficiently.
    This type of inverters is not designed for off-grid applications.

*   **Central inverters - grid-tied;**
    Similar to string inverters but is larger and can connect more strings of solar panels; Strings are connecting to a combiner box which then runs DC power to the central inverter; This type of inverters is not designed for off-grid applications.

Power optimizers:

Devices are located on each panel. Instead of converting DC to AC as microinverters do, they optimize DC power and send it to a string inverter. Designed to minimize the impact of shading on solar system performance and provides panel-level performance monitoring. Can be more affordable than microinverters.

- **BATTERY BASED INVERTERS (OFF-GRID INVERTERS);**
  These inverters use battery storage to convert power. They are battery powered and can be used in off-grid systems. Also, these inverters can be used in grid-tied and grid-interactive systems. This is the type of inverters that we will need for going off the grid or at least reduce dependence on a utility company.

  Off-grid inverter can't sell produced power to the grid.

  On the market is available inverters that have a built-in AC charger that can charge batteries from the grid if solar panels didn't charge batteries fully. But for off-grid inverters, AC connection is one-directional, they can only take power from the grid and can't send it back.
  Also, you can connect a power generator to the AC input of these inverters and you will be able to charge batteries this way. So you can use them in the winter or very cloudy days when your panels can't produce enough power.
  It is great to have these options available and it makes your system more flexible.
When you chose an off-grid inverter you have to be sure that the chosen inverter can handle the maximum wattage load when all your appliances are running and surge load capability for some appliances, like a well pump (when starting can have 3 times higher load). So we have to know and check this rating on the inverter so that our appliances could run smoothly.

Some other features that battery-based inverters can have:
- Remote control (control inverter from the living room or via the internet);
- Display;
- Inverter transfer switch
  Some investors can automatically use a generator or a grid AC load to help the invertor with high loads if the invertor can't handle the load by using batteries;
- Automatic generator start or AC charging (allows automatically start the generator and charge the battery when a charge is low and turning off the generator if they are charged)
- Stacking ability (inverters can be stacked to increase voltage and/or amps). Allows use of multiple inverters and automatically turn-on inverters as needed.

Hybrid inverters (all in one grid-tie inverter) or hybrid grid-tied inverters or battery-based inverters which combines a solar inverter and a battery inverter;

You can par solar panels with a hybrid inverter and it can function as both inverter for the battery bank and inverter for solar panels.

You can install hybrid inverters without batteries, and add them later in the future.

So with these inverters, you can convert the power directly from the solar panels and consume the power with your appliances, and/or send it to the grid (or charge battery bank). Used in stationary solar power systems.

- **Microinverters – grid-tie inverters;**
  Each panel has its own solar inverter that converts DC to AC. This helps to control each panel's performance and each panel is independent in the system. This invertor is not designed for off-grid applications.
  These investors have higher efficiency but they are more expensive to build the system.

Microinverters – 1 tiny inverter per panel.

Advantages:

1) All panels work independently, if one panel gets dirty or shaded - other panels will work just fine. If you will use a standard inverter and connect all panels in series then if one panel gets shaded all other panels stop working efficiently;

2) Low voltage and amps (don't have to buy more expensive large wires when you connect wires in parallel and amps will increase). Don't have to worry that large voltage can cause you a fire;

3) You can monitor each solar panel individually.

The big disadvantage of microinverters that it adds up from 15 to 20% to the cost of a solar system.

If you don't want to install microinverters then you can install a normal inverter and use **DC optimizers** for each panel, to optimize them. Also, about a 15% increase in the cost of a solar power system.

Inverters also can be:
- **Square wave inverters**
- **Modified Sine Wave**
- **Pure Sine Wave or True Sine Wave Inverters;**

- **Square wave Inverter**. This type of inverter derives its name from the type of wave it generates. When visualized, the wave it makes look like squares as opposed to curves.

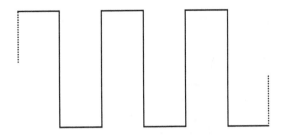

Source: solar-electric.com

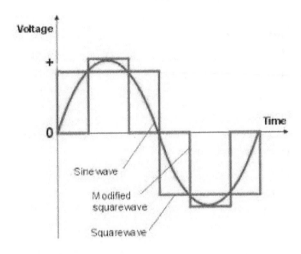

Source: busconversionmagazine.com

Square wave inverters are good at powering simple tools with universal motors. But I don't recommend it to use with other things.

This type of inverters is rarely used nowadays and is not the safest choice for your appliances. Many household appliances get heated when is power by a square wave inverter, and some European countries have banned them. Also, they produce a lot of noise when working.

I don't recommend buying and using these inverters.

- **Pure Sine Wave Inverters**. These types of inverters are the most efficient of their kind. The reason for this is that they bear the closest resemblance to the pure AC current.

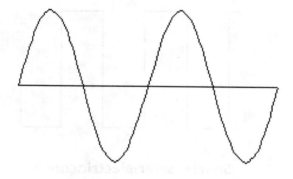

Source: solar-electric.com

Since the pure sine wave inverter is the one that closest resembles the pure AC current wave, it is a lot more efficient and is safe to use for your appliances.

The biggest drawback of pure sine inverters is that they are expensive. So, you can expect to pay a lot more for this type of inverter as opposed to a square wave, or modified square wave inverter.

They produce little noise and provide improved security for major appliances as well as home electronics (monitors, laptops).

You can use this type of inverter with any kind of device, but especially if you have a considerable amount of electronics such as cell phones, computers, stereos, televisions, and so on. They are highly efficient and will help you maintain a consistent load throughout your system.

Also, inductive loads will run more efficiently compared to other types of inverters.

Personally, I feel the safety that they provide justifies the cost.

- **Modified Sine Wave Inverter**. This type of inverter seems to be somewhere in the middle of a square wave inverter and a pure sine wave inverter.

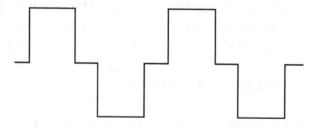

Source: solar-electric.com

The reason for their existence is due to cost. Since there are folks who cannot afford a pure sine wave inverter but want something better than a square wave one, they can choose a modified sine wave inverter.

The justification for using a modified sine wave inverter would be to get more efficiency out of your solar power system without spending a great deal on an inverter. These types of inverters work much better than square wave inverters but not as efficiently and save for some appliances as pure sine wave inverters.

Appliances that work on modified sine wave inverters tend to overuse power, cause additional heating, and because of that inefficiently.

Personally, modified sine wave inverters are a good option for smaller systems such as in a small home or a cabin. If you are looking to power a tiny home or an RV on pure solar power, then a modified sine wave inverter may be a good option for you.

At the end of the day, I would not advise you to purchase a modified sine wave inverter if you can afford a pure sine wave inverter.

There is a list of appliances that you can't use with a modified sine wave inverter. For example medical equipment, laser printers, photocopiers, some laptops, devices with microprocessor controllers, home automation systems, and more.

So the best choice is a pure sine wave inverters.

Also, inverters can be:

•**MPPT Charger Inverters** (that have built-in MPPT charge controller, which we know is the most efficient);

•**PWM Charger Inverters** (have built-in PWM charge controller);

Calculating the size of an inverter

Now that we have discussed the various types of inverters, we need to decide what kind of specs we need on the inverter.

Once again, this depends on the max load you are running on your system.

We can calculate the size of an inverter needed for our system just by summing all appliances load (watts) that we will be using at once:

For example:

Let's assume you are powering a few lightbulbs (50 watts), a television (100), a refrigerator (300 watts), charging phones and laptops (100 watts), and running a fan (50 watts), microwave (700 watts).

Now let's sum up:

Pmax = 50+100+300+100+50+700= 1300 watts

Now we will have to determine our off-grid inverter capacity;

If there would not be any power losses during the conversion of DC power to AC power our calculated maximum load could be the inverter capacity;

But since we know that no power conversion isn't 100% efficient (there are losses on heating), we have to know how efficient is an inverter that we will use;

Let's assume that the efficiency of our inverter is 90%.

So **we can calculate the required solar inverter power rating or capacity this way:**

Pmax / Inverter efficiency

1300 watts / 0.90 = 1444,4 watts;

In this case, we will have to buy at least a 1500 watts inverter since it is not available for sale inverters with 1444 watts capacity.

Whether you choose a pure sine wave, or modified sine wave inverter, you need to make sure that the inverter can handle the max load for your system (1300 watts).

In this example, the max load is 1300 watts. As a result, you will need to use a 1500-watt inverter.

Seems pretty straightforward, doesn't it?

Bear in mind that a pure sine wave inverter may produce somewhere around 95% of the total max load. So, if you are producing 100 watts, the net power that entering your system will be 95 watts. If you are using a square wave inverter, then you might expect to be getting around 90% of your total output. While that might not seem like much on the surface, you might run into trouble when you are going full blast on the system.

So, if your max load is 1300, it is always better to choose at least a 10 or even 20% bigger inverter just to be sure. In our case, at least 1444 watts. Available for sale are 1500-watt inverters.

One other important consideration. If you are running a 12-volt system, then you need to take this into account when looking at the specs of the inverter you choose. So, given a 12-volt system, you might be looking at a 12-volt DC to 230/220 volts / 50 Hz inverter (If you live in Europe) or 120 volts / 60 Hz inverter (USA).

Bear in mind that you need to make sure that the inverter relates logically to the voltage of your batteries. If you are running 24-volt or even 48-volt systems, then your inverter should reflect this voltage.

Some appliances, such as hairdryers, vacuum cleaners, washing machines, and so on, draw much more power before settling down to their normal power consumption. So, this is something that you need to count when calculating your max load. Check a start-up wattage for these appliances and use these numbers to calculate your max load, it happened that you will need an inverter with a much higher capacity.

Choosing an Inverter

For most mobile solar power systems we will require at least a 2000 watt battery inverter.

For stationary solar power systems, it will be much more than 2000 watt inverters (because we might want to use a lot of appliances at once).

Almost all inductive loads like hoovers, motors, microwaves initially required a lot more power than working power to start work. For example, if you use a 1200 watt inverter to power a 1200 watt hoover it will most likely not work, because the initial power that the hoover needs is about 2000 watts. So, for running a hoover you will probably need at least a 2000 watt inverter.

If we choose a too small inverter, it can fail by working too hard to power appliances we will use constantly.

We need to choose between 2 types of inverters:
- MSW(modified sine wave) inverters (cheaper)
- TSW (true sine inverters) or often called a pure sine inverter;

TSW is safer to use with sensitive appliances such as monitors or computers, produces less noise, MSW produces buzzing when working, and with audio equipment can buzzing even more.

Inductive loads run more efficiently when you power them with a pure sine wave inverter.

If you can afford to get a pure sine inverter then get one;

How to find a good one?

You can search on google for top off-grid inverters, also go on Amazon and find some pure sine wave investors with a high rating and positive reviews and with a good price. You will have to do some research here.

Summary

At this point, we have gone over the four main components of your solar power system:
1. The solar panels
2. The batteries
3. The charge controller
4. The inverter

So, we are ready to see the entire system in action. To do so, let's suppose a rather large system that we are going to install.

First, we calculated our load and determined that we will have a maximum load of 1300 watts.

Finally, the choice for this system would be a 1.5-kilowatt pure sine wave inverter. This capacity would allow the inverter to safely handle the load while managing the output from the batteries rather easily.

We can visualize our set up in the following manner take a look at the system again:

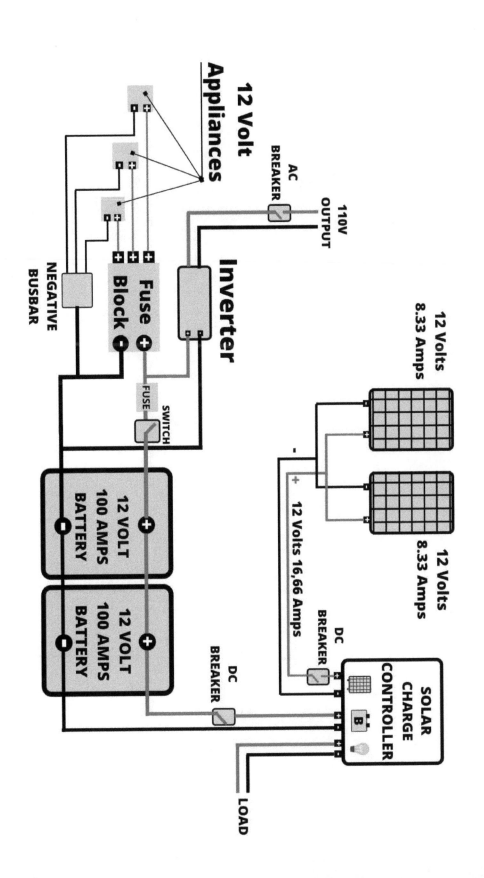

As you can see, the two panels feed into the charge controller. Then, the charge controller feeds into the batteries. Next, the batteries are connected in parallel and feed the inverter.

Finally, the inverter can lead to the main breaker box that distributes the power to the entire home/vehicle.

Your choice of an inverter is just as crucial as the other components of your solar power system. Of course, budgets play a big role in the final decision of what hardware to acquire.

Now, if you don't have a tight budget, then definitely go for the pure sine wave inverter. You will find that you can get much better results. Notwithstanding, acquiring good hardware from a reputable company will provide you with the stability you need in order to make your system work well. As long as you follow the technical specifications of all the hardware your purchase, you should have a fully functioning system.

If you are on a tight budget, I would advise you to shop around and get several opinions before making your final choice. Often, you can find some better deals if you go online. By doing this, you will be able to make the most out of your investment.

At this point, we have come to an end with the description of the various components of your solar power system. In the following chapters, we will drill down into how you can actually build your solar power system. This entails the actual mounting of panels, installation of batteries and solar charge controller, and so on.

Consequently, I would encourage you to take some time and go around your home or an RV. Have a good look at the places you feel would be most suited for the installation of the various components of your solar power system (battery bank, solar panels, inverter, and so on). This will allow you to have a visual of what we intend to describe in the following chapters.

In doing so, you will be able to have a clear vision of how you can set up your solar power system. This will hopefully save you time and effort since you will be aware of how you can take advantage of the physical spaces you have.

So, take the time to do a little research. You will find that the old axiom that says, "for every minute you spend planning, you will save ten minutes in execution" is true.

Chapter 6: Solar panel mounting and wiring

Dimensions of the panels

The first item to consider is the dimensions of the panels themselves. In essence, the greater the capacity on the panels, the larger the size. This relates specifically to the number of cells contained in the panel. Thus, more cells mean more surface to capture sunlight.

Given this rule of thumb, you can then proceed to figure out the best spot for your panels. As I mentioned earlier, many folks like to mount their panels on their roofs. This is a good idea when you have a Gable roof. Besides, having a sloping roof pitched to one side would certainly be of benefit.

Of course, given the dimensions of the panels themselves, they may be too big or too heavy for your roof. This is something that you might want to get a second opinion. In this case, the last thing you want to do is to put additional weight on your roof.

Roof mounting may not be the best course of action if you have an older roof that's not in the best of shape. However, roof mounting may be your best choice especially if you don't have much land on which to mount your panels.

You could have smaller panels though having multiple panels may be more of a hassle than a benefit. Also, if you choose to mount your panels on the ground, make sure that they are in a spot where they can get direct sunlight most of the day. Before actually mounting them, monitor the area in which you would like to mount your panels to see if there isn't any shading.

Typically solar panels are about 65x39 inches or 165 x 100 cm depending on the manufacturer and power rating of a panel you will decide to buy.

By having the watt rating and dimension of one solar panel you will be able to see the number of panels you can install on a certain area.

Wiring of solar panels

Wiring in series (chain):

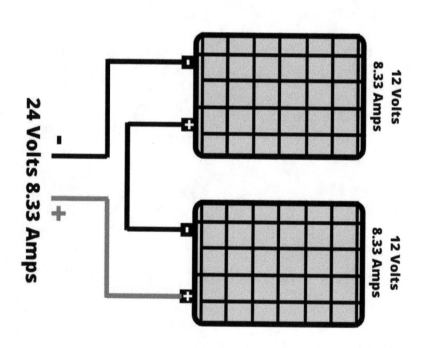

12 Volts
8.33 Amps

12 Volts
8.33 Amps

24 Volts 8.33 Amps

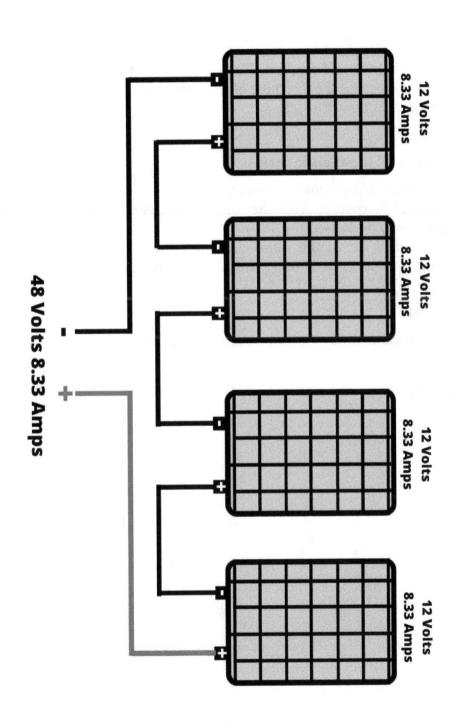

The negative lead from the terminal of 1 panel goes to the positive lead of another, then the negative lead of this panel to the positive lead of the third panel, and so on.

V=V1+V2+V3+V4+...

I=I1=I2=i3=I4

The voltage will increase and amps will be the same. When we have small amps we can choose thinner wire (because the thickness of wire depends on amps going through the wire), and we can have smaller power losses.

Cons of connecting solar panels in series:

1. Solar panels are all connected and work as 1 giant solar panel. If 1 panel of the solar array will be covered from the sun, the efficiency of the hall solar array will be significantly decreased.
2. They should be all the same characteristics (volts, amps, material...) and be mounted close one to another and angled in the same way to work efficiently.

Wiring in parallel

In this case, the voltage will stay the same, but amps will add up.

I = I1 + I2 + I3 + I4

V = V1 = V2 = V3 = V4

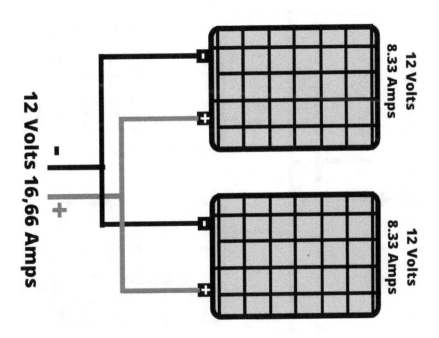

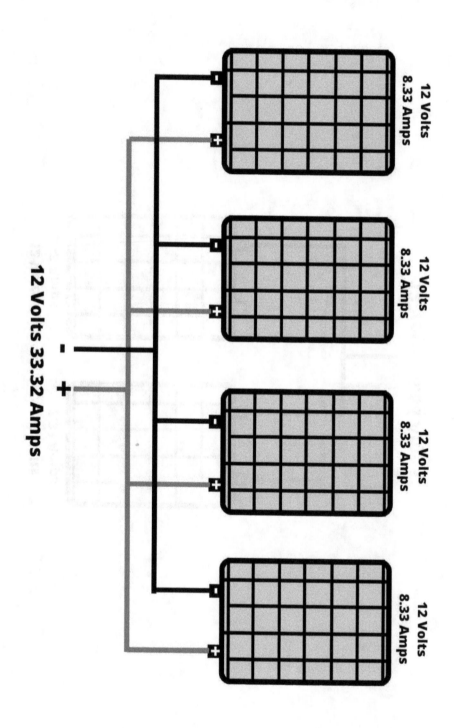

Cons:

- Increased amps will require a thicker and more expensive wire to conduct electricity efficiently;

Pros:

- Panels can have different amp ratings (but the same voltage).
- All solar panels in an array work independently. If one solar panel stop working other panels will steel keep producing electricity well.

Combined wiring:

You can wire 2 or 3 solar panels in series then wire another 2 or 3 solar panels in series, and wire two series groups (strings) in parallel.

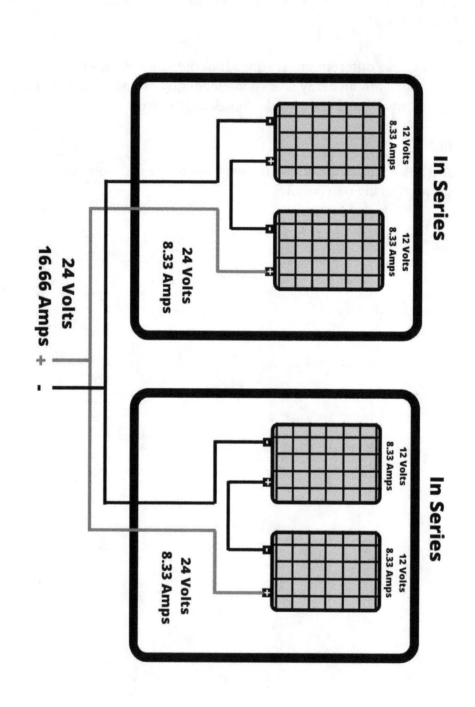

If you have shading problems, for example, you have installed a solar power system on a vehicle and you are traveling (you might park your vehicle in places where sunlight will be shading partially on your roof) in this case efficiency of your system will be lower. But if you will combine solar panels in a series by groups and then wire them in parallel efficiency of the system will be higher when part of your roof will be shaded.

Use MC4 branch connectors to wire in parallel multiple in series groups to make 2 or more wires of the same polarity in 1 wire. When you wiring a lot of solar panels together use zip ties to tidy up cables.

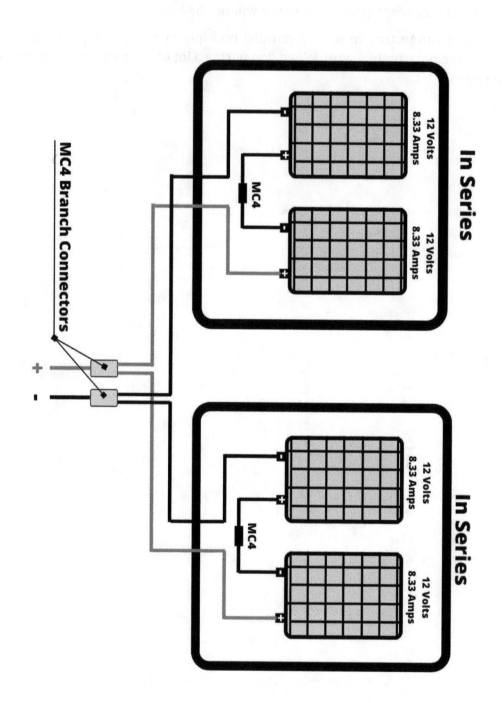

Which way to face

The sun moves along the equator, there is a predictable pattern in which sunlight will travel. Besides, the curvature of the Earth will not distribute sunlight evenly. Sunlight will be distributed in a specific direction, given your geographical location.

For instance, if you live in the northern hemisphere, then your panels should face south. If you live in the southern hemisphere, then your panels should face north. Unless you are living in the exact equator, you can simply put your panels at 10-15 degrees on the ground, to make sure the rain will clean panels. The sun on the equator is high all year.

If you are unsure about which direction is north and which is south, you can look at a map tool such as Google Maps, or you can use your car's GPS. Those are two very simple ways in which you can determine your north/south position. Or you can use a compass.

The reason for facing panels in the opposite direction of your hemisphere is related to the way the Earth itself is tilted. The Earth tilted axis is the reason why we have seasons. As such, when it is winter in the northern hemisphere, it is summer in the southern hemisphere and vice-versa. The only part of the world, which gets an equal amount of sunlight throughout the year, is the exact equator.

Tilting solar panels

The best angle to tilt your panels is – the latitude.

Ideally, your panels should be tilted perpendicularly to the sun - at an angle, so that we can get the most sunshine and as a result high power output.

The latitude is changing depending on where you live.

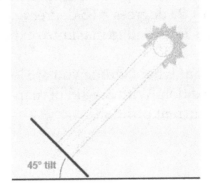

45° tilt

Source: eia.gov

The sun is moving in the sky and can be lower or higher depending on the time of the day and the season of the year.

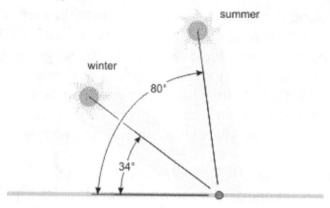

The latitude and angle of tilt for panels will be bigger the further you live from the equator.

There are few ways that you can calculate your tilt angle.

Example – When your latitude is 40 degrees;

Method #1 (Simple)

Add 15 degrees to your latitude during the winter, and distract 15 degrees to your latitude during summer.

Summer: 40 -15 = 25 degrees;

Winter: 40 + 15 = 55 degrees;

Method #2 (Optimal – better results)

Summer: (40 degrees x 0,9) - 23,5 degrees = 12,5 degrees;

Winter: (40 degrees x 0,9) + 29 degrees = 65 degrees;

The previous calculation is intended to maximize exposure to sunlight during daylight hours.

In case you are unsure about what latitude you are located at, don't worry. You can search for your geographical position. With the aid of map tools such as Google Maps, you can easily determine your current position.

All these calculations make the most sense for utility-scale solar power plant installations. If you plan to install a residential solar power system or a small rooftop system then additional costs that go into tilting your solar panels and maintenance of all construction will not be worth of small increase in power production. So most people just install solar panels on their roof without considering the angle of solar panels. (A few kilowatt-hours increase in power per year might cost you hundreds if not thousands of dollars/euros). But it is up to you, each situation is unique.

If you have a small off-grid system and each watt of power matters for you, then tilting your panels to the latitude of the location where you live can make sense.

How to mount solar panels

There are two main factors that you have to know to mount your solar panel:
First where they are going to be;
Second, how much you are willing to spend.
 Mounts for solar panels:
1. Brackets - cheapest and easiest to use, more beginners friendly;
      Downside – is that you not always can allow about 10 cm (3 to 4 inches) of space under panels for proper ventilation and cooling the panels;

2. Pole mounts

Designed to be mounted to a round metal pole. They are more expensive. Used in remote power applications, or if you don't want to mount them on the roof and have only a few panels.

Source: solarelectricsupply.com

3. Ground-mounted rack system

If for some reason you want to place your system on the ground, maybe because you don't have much space on the roof, or if your roof facing in the wrong direction you can use this rack system and mount a lot of panels on the ground. This system is very expensive. To not let them blow over - typically they are anchored in the ground with concrete or any other method.

Source: solaracks.com

4. Rooftop rack mounting systems
They are strong and can hold a lot of panels.
This system is mounted to the roof trusses.

There are some other types of mounting for solar panels. For example, I have found useful some adjustable mounting systems where you can regulate tilt for panels manually.

And there is a mounting system that can automatically track and follow the sun to ensure maximum power output, but this system is extremely expensive and is used by solar power plants. But you can buy or build them yourself if you want.

First what comes to mind is to use a solar panel mounting kit and attach solar panels directly to the roof by L-brackets and long bolts and washers after this using caulk to seal all of the holes in the roof.

But not all surfaces are good for this, and having holes in a vehicle isn't a good thing to have. Most vehicle roofs are not designed to have solar panels mounted on them and are made from fiberglass and foam, and L-brackets should be bolted to metal or wood.

For homes, it is not a problem. And solar panels are mounted to the roof this way safely.

Another option that you might have is to **bolt solar panels right to a roof rack.** It depends on the vehicle that you have.

If you decide to use a flexible solar panel because of a curved roof or because you can't add a roof rack and your roof is made from fiberglass then your only option might be to use a mounting tape that a solar panel manufacturer recommends.

Also, you can use "a drill-free corner mount solar panel brackets" and a VHB tape.

Make sure that panels are secured well and not fly away while you drive. Always remember about safety. You can add and probably have to add a safety line to the solar panel array mounted on a vehicle. Drill holes in solar panels frames, and use a stainless steel cable or a rope to attach all solar panels together with a safety line, and attach a safety line to a roof rack, or any other sturdy object.

Roofing Solar Panels

1. First layout where the perimeter will be for your solar panels, with a lumber crayon and chalk line. This will help to visualize where panels are going to be and help to lay racks on top of which are going to be our solar panels.

2. If you will mount your solar panels statically to the roof, and your solar panels came with mounting brackets. The first thing we should do is mount the mounting brackets to the solar panels (a bolt, a washer, a lock washer, a nut). First a bolt and a washer, and on the other side a lock washer and a nut. Then we can screw the solar panels to the roof of a hose or a vehicle.

3. To attach the racks to a roof we need to identify where are the trusses underneath the roof. We can easily do this with a hammer. Listen to the sound from the hammer, you will notice the difference between the loose area and the solid. After this, we can drill down and attach our leg bolts.

4. Never put your weight on the middle or edge of solar panels while installing solar panels. Use rails to mount panels place your weight on the mid-clamps. Banding solar panels when you put weight in the middle of the solar panel makes small cracks in the crystals of the cells decreasing their efficiency.

5. Put roof flashing in places where you drilled the halls in the roof, and use a sealing around the hall in roof flashing and inside the hall of the roof to keep water out of the roof.

6. After this, we have to drill a hole through the roof to run wires from solar panels to a charge controller. We locate this hall underwear of 1 of the panels, we place flashing, and drill the pilot hole from the top down, then we use a hole saw to drill through the roof to the other side.

Grounding

1. We should attaché a system ground on racks. This is a copper wire that tightened to each rack so that if something happened faulting or lighting, this wire will just transfers electricity to the ground.

2. We will need to drove a steel rod with a copper coating in the ground then connect a grounding wire to the steel rod.

The mounting process can be a little bit different depending on the type of mounting system you choose. So you have to read recommendations and explanations for mounting that will come with your mounting system when you buy them.

Chapter 7: Series and parallel connections for batteries

In this chapter, we will be looking at the hook up that needs to be done to build the battery bank.

Hooking up the batteries

There are two ways of hooking up batteries:

      1.   A series connection

      2.   A parallel connection

Both types of connections will get the job done;

So, if you have two 12volt, 100 amp-hours batteries, each of them will storage 1200 watt-hours of energy (12 volts x 100 amp hours = 1200 watt-hours). Hooked up together will have a capacity of 2400 watt-hours (1200 watt-hours x 2 = 2400 watt-hours).

The watt-hour capacity of your system will remain the same regardless of the way you hook up the batteries. What will differ is the voltage and current of the system depending on the setup.

Here is the difference:

- A series connection will add up the voltage of all the batteries and amps will stay the same. So, if you are running two 12-volt batteries, then you must add up the voltage of the system. Thus, you would be running a 24-volt system.

- A parallel connection will not affect the voltage. So, if you have two 12-volt batteries, the batteries will still produce 12 volts. The difference will be visible in the Amperage of the system. Amp-hour rating will add up.

100 amp hours + 100 amp hours = 200 amp hours battery bank.

Which type of connection to choose

Advantages of series connection (voltage add-up, amps don't change):

- You can use smaller wire sizes; Amps will not increase so you can use smaller wire size;

- With with increase of voltage – increase solar charge controller energy output (while solar charge controller amperage stays the same).

> Your controller just should be rated for the voltage of the system you have.

> If you connect batteries in parallel you will have to buy a solar charge controller with high amps output (which will be more expensive);

Advantages of parallel connection (amps add up, the voltage don't change):

- You can build a safer system. 12 and 24 volt systems are safer – you will not get an electric shock with smaller voltage systems like 12 volts and 24 volts systems (everything about 30 volts technically may shock you);

- You can plug into the 12-volt battery bank a 12 volts DC appliances without using DC/DC converters to decrease voltage to 12 volts;

The disadvantage of the parallel connection:

- When amperage will go up you will have to increase wire sizes;

To get 12 volts from the 48-volts battery bank you will need to use a DC/DC converter that is easy to get and is affordable.

Leaving the system at 12 volts would be best for small systems.

You have to match voltage (12 volts, 24 volts, 48 volts) for a solar array, battery bank, and inverter.

How to decide voltage for your battery bank?

If you plan to power only 12-volt appliances in your RV or boat... then obviously you will need to build 12 volt battery bank.

If you plan to power an AC appliance like a fridge or TV in your hose, for this you will need an inverter, then what is the voltage of that inverter?

In most cases the higher the output wattage of an inverter the higher the DC voltage input.

Up to 1500 watt inverter will probably require 12 volt DC input;

5 000 to 10 000 watts inverter probably will require 48 volts DC input;

How to get high performance for battery set up:

- When you connect batteries in 2 or more groups of batteries that are wired in series and then connect those groups in parallel – always make groups the same (mirror them) – this will ensure the equal performance of all batteries;

- Use the same batteries, batteries age, the same connectors, wire length, and wire size;

- All batteries should have the same amp-hours, the same voltage;

- Install batteries in the same temperature zone, and keep the optimal temperature for your batteries. Do not allow your battery bank to heat up – allow proper ventilation;

Also, if you have a bigger system – for example, your 5000-watt inverter requires 48 volts to run then you need to build 48 volts battery bank (by connecting them in series) to match the voltage of the inverter.

You can use a combination of both wiring types to get any voltage and amps you want for your battery bank.

Hooking up at the series connection

In this type of connection, what you are doing is creating one giant battery. So, instead of having two individual batteries, you have one large battery made up of two components. In this way, you are adding up the voltage of all batteries while keeping the Amp-hour rating constant.

Hence, if you have four 12-volt batteries, then you would be running 48 volts. If you have two 24-volt batteries, then you would also be delivering 48 volts.

In essence, the way to hook up a series connection is as follows:

- **First, we will have to wire all batteries together** connect the negative terminal of the first battery to the positive terminal of another battery, then connect the negative terminal of the second battery to the positive terminal of the third battery, and so on - as is shown on **Figure 3** (next page).

- **Second, we will connect the wired batteries to the charge controller** (only after this connect charge controller and solar panels). Positive lead to positive (where shown the battery), and negative lead to negative (read instructions for your solar charge controller to be sure, and to not damage it).

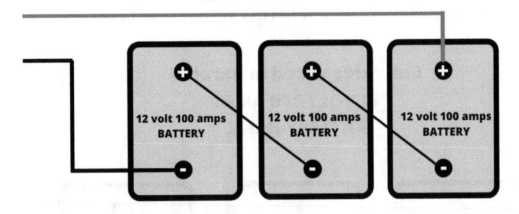

Batteries wired in series
36 VOLTS 100 AMPS
BATTERY BANK

As you can see in the diagram, the negative terminal of the first battery is connected to the positive terminal of the second battery... Then, the positive terminal of the first battery goes out to the solar charge controller while the negative terminal of the second battery also goes out to the controller.

In this case, you are adding up the voltage of 3 batteries. Thus, you are running 36 volts while maintaining your Amp-hour rating constant.

If your batteries don't have a BMS - overcurrent protection, then you will have to add a fuse on the positive lead that will lead to inverter;

Setting up a parallel connection

The system would be set up as follows:

- The positive terminal of the first battery would feed into the positive terminal of the second battery, and so on. The negative terminal of the first battery would fit into the negative terminal of the second battery and so on. **(Figure 4 – next page)**
- Next, we connect the batteries to a solar charge controller.
- Then the charge controller to solar panels.
- And only after this, we connect the batteries to the inverter (negative lead from the batteries to the negative terminal of the inverter, and then the positive lead from batteries to the positive terminal of the inverter).

The increase is not in voltage but in the amp-hour rating, so you would need to make sure the voltage rating of the panels, solar charge controller, and the batteries match up. So, if you are running 24-volt panels, then make sure you have 24-volt batteries. That way, you can be sure that you won't be overcharging your batteries.

The following diagram illustrates the way this type of connection works:

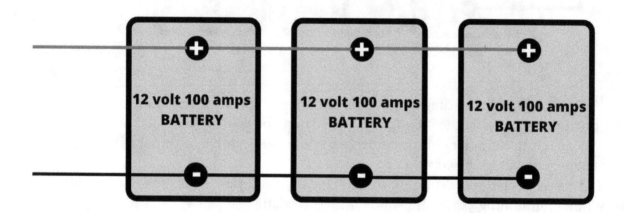

Batteries wired in parallel
12 VOLT 300 AMPS
BATTERY BANK

In the figure above, you can see how the negative terminal of both batteries (black) is connected directly to each other. There is a wire that leads from both terminals and feeds into the inverter. Likewise, both the positive terminals (orange) are connected together with a wire running from both wires.

Combined wiring – 48 volts battery bank

12 Volt, 100 Amp Batteries wired in combination
24 VOLTS 200 AMP HOURS BATTERY BANK
CREATING 2 SERIES GROUPS

12 volts + 12 volts + 12 Voltd + 12 Volts = 48 volts
100 amps = 100 amps = 100 amps = 100 amps

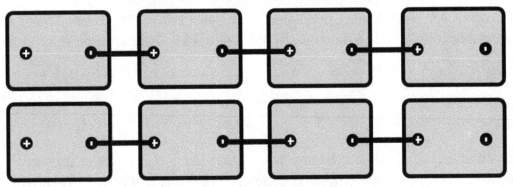

12 volts + 12 volts + 12 Voltd + 12 Volts = 48 volts
100 amps = 100 amps = 100 amps = 100 amps

Conecting 2 Series Groups in Paralel

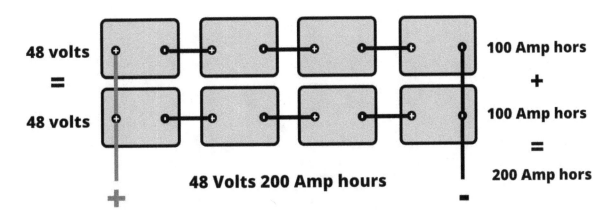

48 volts

=

48 volts

48 Volts 200 Amp hours

100 Amp hors

+

100 Amp hors

=

200 Amp hors

+ −

Connect a bolt-on fuse on a positive lead of a battery bank, and tighten the battery terminal bolt with a wrench.

It is recommended to isolate positive terminals of all batteries to avoid creating short circuits between positive and negative terminals by electrical tape.

One word of caution: it is important to make sure that the inverter is equipped to handle the level of Amperage being produced by the system. The Amperage level may lead the inverter to overload and damage. In the best of cases, the inverter may blow a fuse.

Summary

In this chapter, we discussed the two ways in which you can power up your system. As such, you need to take great care in running tests prior to fully making your system operational. If you happen to make a mistake by wiring batteries backward, you could cause a short circuit. Also, you may end up damaging your batteries. So, it certainly pays to draw up a schematic of the way you will hook up your system, follow that schematic, and then set up the system.

Chapter 8: Setting up equipment and wiring the system

At the outset of this book, I encouraged you to walk through your home and decide where you want to place your hardware. As a general rule of thumb, the best place to install the batteries, the charge controller, and the inverter would be in the same location as your main breaker box. In doing so, you will have access to all of the hardware needed to power your home or vehicle in the same spot while minimizing the wire losses between them.

1. **First, we place batteries**
2. **Connect the solar charge controller to batteries**
3. **Solar panels to charge controller**

Where to put the batteries

We will idealistically require a place:

- Insulated from large temperature fluctuation.
- Where the battery can be secured (with a strap) so that it does not tip over (if it is a vehicle).
- A dry place that is protected from moisture.
- Ventilation if you don't use a sealed battery.
- A compartment should be somewhere in the middle of a vehicle between the front and back parts of the vehicle (where is located center of gravity of a vehicle).

One other important safety tip: To reduce the likelihood of your batteries' terminals coming into contact with any other substance and potentially setting off a fire, you can wrap the positive terminals with a large, flat piece of rubber or at least insulating tape. Rubber doesn't conduct electricity, and it is very good at isolating sparks and such. So, keep this in mind just as a safety precaution.

What about the charge controller?

The charge controller is a small piece of hardware that doesn't weigh very much. So, you can easily mount this on the wall. Preferably, you could mount it about chest-high. That way, you can easily see it and read it.

Think of the placement for your thermostat. You can place it about that high. Also, you want to keep your charge controller close to the entire set up. That way, you can check in on your solar power system once in a while, minimize power losses for wires, and make sure that everything is set up the way it is supposed to be set up.

I would not advise you to locate your charge controller inside the house while your entire setup is in the basement. The reason for this is that if you have the charge controller inside the house while the setup is elsewhere, then you won't be taking a look at the system as often. By having the charge controller near the hardware, you will be forcing yourself to take a look at your system every time you check the charge controller.

So, the solar charge controllers should be mounted close to the batteries, typically somewhere on a wall. A charge controller has cooling fins for convective ventilation. So make sure that under and above a charge controller is some space for this.

How to mount? Just use some screws or mounting tape if your charge controller is lightweight, but not recommended.

Where to place inverter?

The inverter has to be placed close to the battery bank so that has minimum power losses for wires. Typically you can mount an inverter on a wall.

Choosing wires

The efficiency of the system largely depends on the wires (material that the conductor is made from and thickness) you have used in the system. the copper conductor has high conductivity.

Gage size is the overall thickness of the wire. Wire gage size for solar power systems ranges from the thinnest 14 gauge to the thickest 0/4 gauge wire.

The thickness of the wire dependent on the length of the wire, and amp load that wire should carry;

If you choose smaller wires that are required you can cause overheating, which can lead to fire and huge power losses in the system. If you chose the wrong wire even a fuse will not save you.

Recommendations:
- Always choose a little bit thicker wire then required if it is possible;
- For wiring solar panels use a solar hook up wire (UV (ultraviolet) resistant);
- For long distances use thicker wire to conduct current efficiently;

If you don't want to use thick wires (copper) because they are very expensive or because can't find one, then you can increase the voltage for that wire just by connecting elements of the system (batteries or solar panels) in series.

Wire size recommendations:

- **If you wire 12-volt panels:**
  10 gauge wire if the length is less than 25 feet;
  8 gauge wire if the length is more than 25 feet;
  If you connect 2 12 volt solar panels in series you will have 24 volts total, now you can use 12 or 10 gauge wires.
- **Wiring a solar charge controller (amps) and battery bank:**

  20 amps – 12 gauge
  30 amps – 10 gauge
  40 amps – 8 gauge
  60 amps – 6 gauge
  80 amps – 4 gauge

Choosing wire for wiring an inverter and battery bank:

Inverter carries a big current, and require very thick wire. If you don't want to look for wires and installing fuse and connectors. You can buy an inverter wiring kit online with an already installed fuse.
1000-watt inverter – 4 gauge wire
2000-watt inverter – 0 gauge wire
2500-watt inverter – 2/0 gauge wire
3000-watt inverter – 4/0 gauge wire
For running appliances you can use 12 gauge or 10 gauge wire safely;

12-volt wire gauge chart
Source - qualitymobilevideo.com

Amperes

Amperes	Length in Feet						
	0-4 ft	4-7 ft	7-10 ft	10-13 ft	13-16 ft	16-19 ft	19-22 ft
250-300	4 gauge	2 gauge	2 gauge	1/0 gauge	1/0 gauge	1/0 gauge	2/0 gauge
200-250	4 gauge	4 gauge	2 gauge	2 gauge	1/0 gauge	1/0 gauge	1/0 gauge
150-200	6 or 4 ga.	4 gauge	4 gauge	2 gauge	2 gauge	1/0 gauge	1/0 gauge
125-150	8 gauge	6 or 4 ga.	4 gauge	4 gauge	2 gauge	2 gauge	2 gauge
105-125	8 gauge	8 gauge	6 or 4 ga.	4 gauge	4 gauge	4 gauge	2 gauge
85-105	8 gauge	8 gauge	6 or 4 ga.	4 gauge	4 gauge	4 gauge	4 gauge
65-85	10 gauge	10 gauge	8 gauge	8 gauge	6 or 4 ga.	6 or 4 ga.	4 gauge
50-65	10 gauge	10 gauge	10 gauge	8 gauge	8 gauge	8 gauge	6 or 4 ga.
35-50	10 gauge	10 gauge	10 gauge	10 gauge	8 gauge	8 gauge	8 gauge
20-35	12 gauge	10 gauge	10 gauge	10 gauge	10 gauge	10 gauge	10 gauge
0-20	12 gauge	12 gauge	12 gauge	12 gauge	10 gauge	10 gauge	10 gauge

Wiring the system

1. Wiring solar panels together into a solar array
2. Wiring the batteries
3. Connecting the solar charge controller to a battery bank
4. Connecting solar panel array to the solar charge controller
5. Connecting an inverter
6. Connecting a DC fuse block
7. Installing a battery monitor or voltmeter

Wiring solar panels and batteries we already discussed in previous chapters so now we will need to connect the solar charge controller and battery bank.

Connecting a solar charge controller to the battery bank

We have to connect the solar charge controller to the battery bank first because there is a risk that you will damage the solar charge controller (electronic) if connect solar panels first.
We should wire a solar charge controller and batteries and after the solar charge controller and solar panels.
First, we connect a wire to the negative terminal of the battery then to a negative terminal of the solar charge controller (should be marked as a battery). For this, we have to strip one end of the wire and attach a crimp connector that will be connected to the negative terminal of the battery, after this another end of the wire we have to strip and connect to the solar charge controller (by using a screwdriver or Allen wrench).
Then, strip and insert a wire to the positive terminal of the solar charge controller and run the wire to the battery bank and attach it to the battery bank fuse by using a crimp connector.
If you have made everything right – you will see a green light on the solar charge controller showing that the batteries are connected.
After this, we can start wiring solar panels and connect them to the solar charge controller.

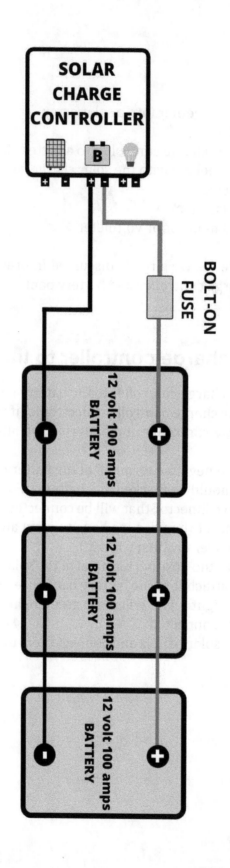

Connecting solar panels to a solar charge controller

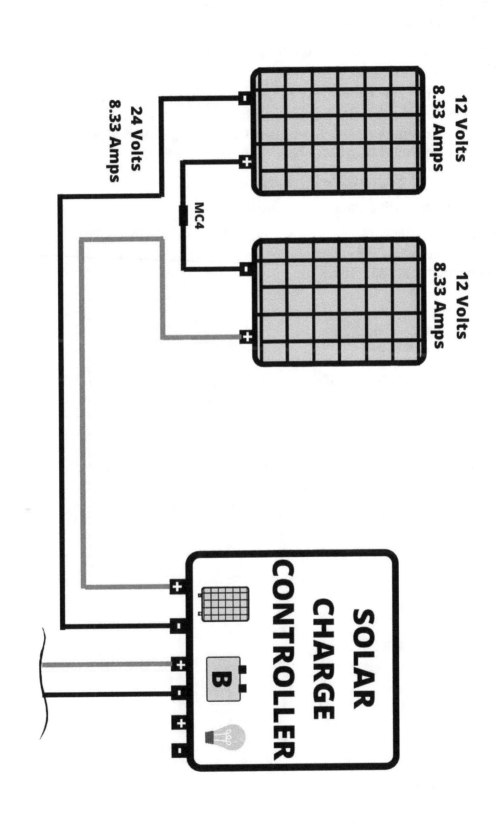

For this, we have to pass wires through the roof. In most cases, you have to drill 2 holes into the roof and later to seal the holes with caulk.

A lot of people use "a cable entry gland" and self-leveling sealant to make it look good and seal holes in the roof.

Before connecting the solar panel array to the solar charge controller when you connected a lot of panels in series, check if the voltage isn't higher than the maximum voltage of the solar charge controller to not burn it. To measure the voltage of the solar array by using a multimeter, and also would be great to check the polarity of wires.

All MC4 connectors should be located on the roof.

How to connect wires in the solar charge controller?

Read the manual of your solar charge controller. Positive wire to positive sign "+" on solar charge controller, negative "-" to negative. When everything is made right you will see a green light that shows that solar panels are connected and charging batteries.

Connecting Inverter

The inverter will supply 120 volts (if you live in the USA) or 220-240-volts if you live in Europe (Germany and France - 230 volts).

Connect the inverter to the battery bank directly using an inverter wiring kit (that might come with an inverter, or buy it separately), or choose the wire thickness the way I described above. Bolt positive cable to the bolt-on fuse, and after this connect the negative cable. Usually, the negative terminal (the wire that will be connected last) will spark when you connect the inverter and it is normal. Now you can start the inverter and run the AC appliances by using the inverter.

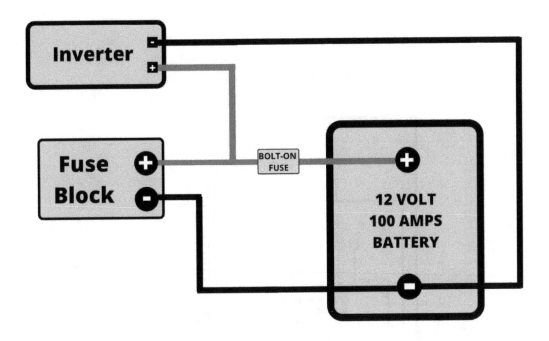

Connecting the fuse block

The fuse block will supply 12 volt DC appliances. If you will have a higher voltage battery bank (24, 48 volts) then you will have to buy the converter to get 12-volt power, as I said before.

We require a 2 to 8 gauge wire, to see what is recommended for your fuse block, read the manual. Or chose the cable base on the maximum current (amps) that the fuse block will have to provide.

Mount your fuse block close to the batteries. A fuse block has a positive and negative terminal which we need to connect to the battery bank.

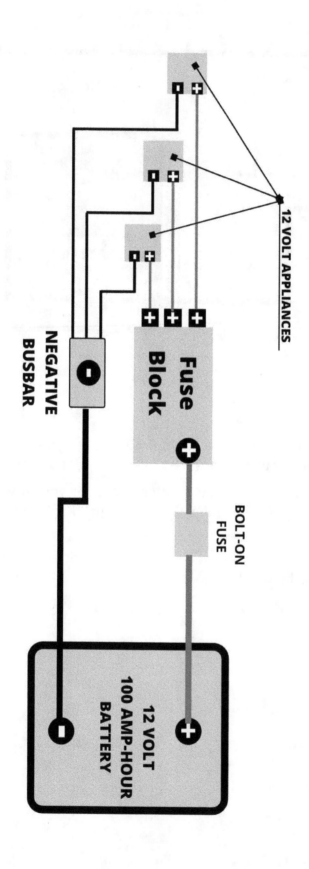

How to connect a battery monitor and why we need one?

Increase the life of the battery (by charging it to 50 or at least 70% for a lead-acid battery, and 20 – 50% for lithium batteries).

We can estimate the health of the battery, simply knowing how much power a battery is storing and estimate how long the battery will last.

We need one to know how much power we have drawn out of the batteries and figure out a true state of charge.

One of the most popular and best battery monitors are shunt. There is also the battery monitors that have a hole sensor.

Most battery monitors are showing the battery capacity, the voltage of the battery, amps, what-hours, watts, and more. More expensive battery monitors can show volts, amps that going in and out of the battery.

Most battery monitors provide instructions, and a wiring diagram for battery monitor so that you couldn't mess it up.

A shunt has a post on each side and two screws in the middle. The battery monitor will be connected to those 2 screws and the monitor showing the current that goes through this shunt.

Depending on where you wire your shunt, this will determine what a shunt will measure.

If you want to measure how much power your solar panels are producing (voltage, amps) then connect the shunt before the solar charge controller.

For batteries, disconnect a negative wire that is connected to the battery (but first disconnect one of the main solar panel wires from the solar charge controller to not damage the charge controller) and connect it to one end of the shunt, and add one single wire that will go from the second end of the shunt to the negative terminal of the battery. A lot of shunts can be bolted to the negative terminal of the battery. Everything should go through this shunt so that we could measure total electricity. After this, we wire battery monitor wires to the interior screws of the shunt and battery monitor.

To give the power to the monitor we connect positive and negative leads to the battery, as is shown on your wiring diagram that comes with the battery monitor. Each shunt will be different.

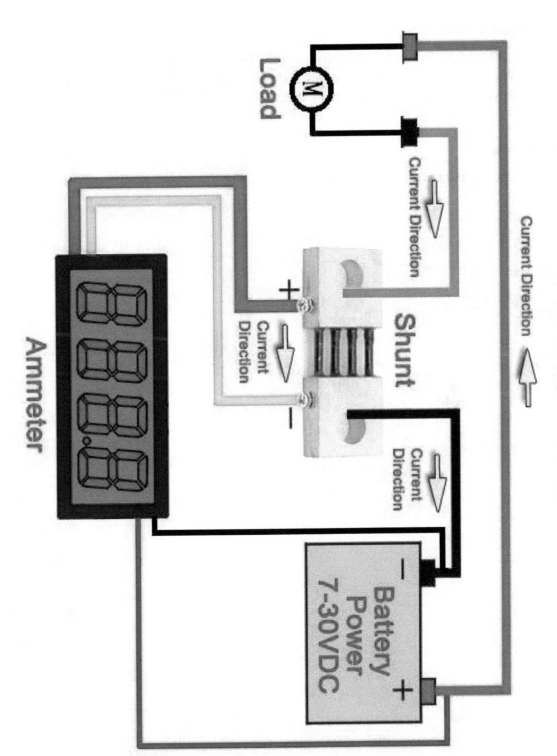

Wire Connection

Load

Ammeter

Shunt

Current Direction

Current Direction

Current Direction

Current Direction

Current Direction

Battery Power 7-30VDC

How to connect a DC 12-volt appliance?

You can wire appliances to male type plug connectors (XT-60 or Anderson Powerpole). For this, you will have to hardwire a female type connector to the fuse block. We can use appliances when we want. Or you can hardwire appliances directly to the fuse block.

XT-60 connectors:

Source of the image - www.batteryspace.com

Male   Female

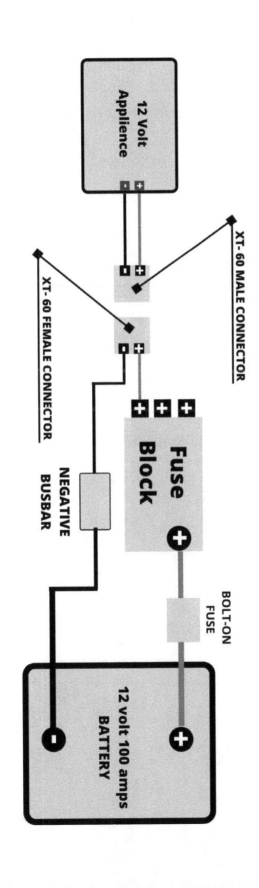

Always make sure that you are using a proper gauge wire and fuse when connecting 12-volt appliances, also only use a 12-volt appliance not smaller, because otherwise, you will damage an appliance.

XT-60 connectors are designed to power up to 60 amps of appliances.

You cant use XT-60 with large appliances. For larger appliances, you have to use an Anderson Powerpole connector.

To use an XT-60 connector, you will need a soldering experience. Recommend to watch some videos online to see and learn to solder and experiment a little bit, before soldering connectors.

This is what your final wiring diagram should look like:

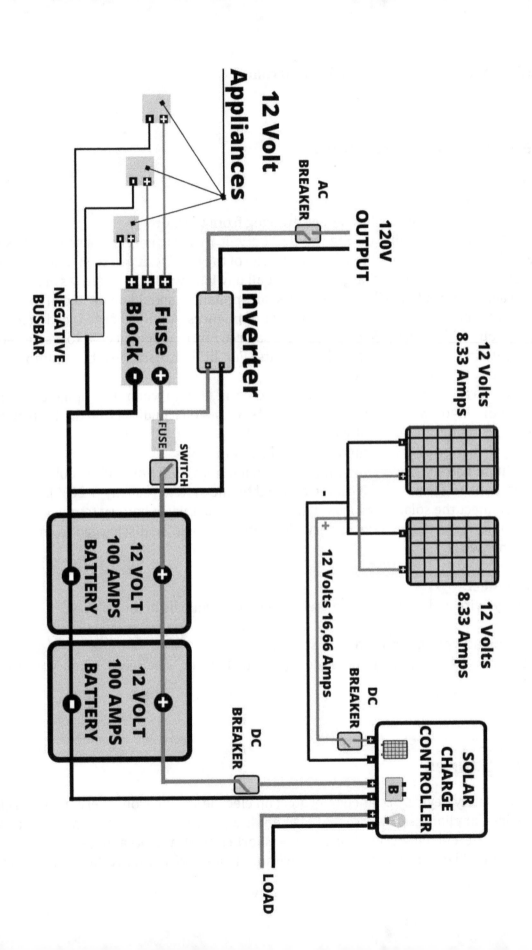

Figure 5. - Solar power system wiring diagram

Battery bank voltage meter

Why use?

Helps to measure the charge of a battery; (volts)

Amount of electricity that appliances are drawing from the battery;

While batteries are charging during the day on volt monitor is present so-called **"surface charge"**. The voltage is higher than the real charge of the battery. It can seem like the batteries are charged up, and a surface charge is present all day long while the batteries are charging.

To see an accurate reading of batteries charge all appliances should be turned off because the voltage on the monitor drops when you power appliances. A fully charged lead-acid battery will have about 14,5 volts while charging by a solar array, if not connected to a solar array or during night time will be about 12,7 -13 volts. For lithium batteries, this number can be different, check the batteries manual.

We can discharge a lead-acid battery up to 50% or 12,1 volts when the appliance load is off. As I said before if you want to increase your battery life you can discharge the battery only to about 70-80% that is around 12,4 volts.

For deep cycle batteries, it is around 20% or about 11,8 volts.

If your lithium battery has BMS (battery management system) you will be able to discharge the battery bank to the minimum charge allowed by the system (usually 0-20%).

If you disconnect the solar array from batteries and don't use a battery to power appliances for a long time, it can damage the battery. The battery should be discharged from 80-90 % at least once per weak.

How to connect:

Use 14 gauge wire and wire voltage meter directly to terminals of the battery;

Where to install voltage monitor:

Mount the monitor in an easy to see location so that you can see it when you are using appliances.

Choosing fuses

A fuse – is a safety device, designed to protect an electric system from overcurrent an electric system (wires, appliances). An essential element is a piece of metal or a strip that is designed to melt when overcurrent occurs (for example short circuit), and breaking an electric circuit.

If overcurrent will occur and the wires are without fuses a wire can overheat and cause a fire.

Fuses are rated in amps – if you know how many amps a wire should carry, you will be able to choose a fuse easily.

Exist 3 main types of fuse holders:
- **Fuse block –** a block of 3 to 10+ (typically from 10 to 30 amps) fuses.
- **In-line fuses** (from 5 amps to 150 + amps) – usually for solar power systems we use from 5 to 20 amp fuses. Much less often 25+ amps fuse.
- **Bolt-on fuse and fuse holders (50 to 250 amps)** – they are designed to attach to the positive lead of a battery bank terminal; We should attach a bolt-on fuse to lead that connects to the inverter and the fuse block or other large appliances.
  This fuse should match to an inverter amp load.

To calculate a fuse size – we will have to multiply a total appliance amp draw by 1,25 (25% more than wire is designed for).

Applience Amp Load x 1,25 = Fuse Amp Rating
If an appliance draws 30 amps, use a 40 amp fuse (30 x 1,25 = 38 amps), or 35 amps if you want to be more on the safe side.

If you have a 3000-watt inverter we will need a 310 amp or 315 amp fuse (3000 watts/12 volts = 250 maximum amp load, then using the formula above 250 amps x 1,25 = 313 amps)
Also, you can choose a fuse rating simply using the fuse size chart below.

Source file: explorist.life

Gauge	Max Amps	Recommended Fuse Size	Maximum Fuse Size
16	20	25	30
14	25	31.25	37.5
12	35	43.75	52.5
10	45	56.25	67.5
8	60	75	90
6	80	100	120
4	120	150	180
2	160	200	240
1	210	262.5	315
0	245	306.25	367.5
1/0	285	356.25	427.5
2/0	330	412.5	495
3/0	385	481.25	577.5
4/0	445	556.25	667.5
BASED ON 105°C WIRE INSULATION			

Choose a fuse size that is below the maximum fuse size or recommended fuse size.

1. First, we will have to attach a large bolt-on fuse between a battery bank and an inverter.
2. After this, we will need to add a fuse (a bolt-on fuse for example) between a solar charge controller and the battery.

This fuse should be 25% larger than a solar charge controller rating.

For example, **if we have a 35 amp solar charge** controller we will need a 45 amp fuse (35 x 1,25 = 44 amps). We will round up to 45 amps (the nearest fuse size rating).

3. Attach a fuse between the battery and a fuse block.

Check the fuse block manual to see what fuse size is recommended.

How to maximize the efficiency of your solar panel system

1. Keep the angle of solar panels equal to the latitude of the place where you live;

2. At least 4 inches or 10 cm space under the panels - to make good ventilation (cooling) for solar panels;

3. Wiring - keep the length of wires between solar panels, solar charge controller, batteries, and inverter - short (the smaller length of the wire and the bigger gauge of the wire the better).

The bigger length of the wires the bigger drop in voltage you will have. The too-small size of the wire will increase energy loss and will cause heating for the wires;

4. Choosing the right solar charge controller. The MPPT charge controller has a lot higher efficiency in comparison to the PWM charge controller.

5. For optimal efficiency, your batteries should be about 77-79 degrees Fahrenheit or 25-26 degrees of Celsium. Check the manual of a battery to know for sure, but typically it is so. Higher the temperature will decrease the life of a battery, and a lower temperature will decrease power storage capacity.

6. Avoid shading for the panels. It can seem obvious but a lot of people don't understand how much it can reduce efficiency. Shading of the panels drastically reduces the power output of the panels starting from 50 up to 80%. If shade covers the bottom 10% of the panel you can reduce efficiency by 80%. Avoid shading at all costs.

7. Build a higher voltage system. Building a 48 volts system can be a lot cheaper and more efficient for big systems like in a house. For example, 24 or 48 volts will have smaller wire losses and you will not need to buy big copper wires that are expensive. You will need to build a 48 volts battery bank and at least 80 volts open circuit panel voltage. You need the voltage of the solar array much higher than the combined voltage of your batteries.

For powering 12-volt appliances you will have to buy a DC/DC converter that will convert 48 or 24 volts to 12 volts and is pretty cheap to buy. 12 and 24 volts systems are safe and you will not get shocked if you touch a wire. For 48 volts it is a little more dangerous, be careful.

Things to keep in mind - installing and running the system

We have now basically reached the end of the entire solar power system installation process. So, I'd like to close things out with some things to keep in mind through the installation process.

- Make sure that the voltage is balanced between batteries and panels. To avoid having to do some extra math, you can purchase batteries that have the same voltage as the panels. That way, you can hook them up on a parallel connection.
- If possible, get an MPPT charge controller. This is the most efficient type of controller. If you cannot spring for one, the PWM charge controller will do a good job.
- Use a pure sine wave inverter whenever possible. This is the best kind of inverter as it simulates the AC wave very closely.
- It is always a good idea to keep your equipped raised off the ground. This is especially true if you plan to set it up in your basement.
- Monitor your system regularly. Once a week seems to be a good timeframe to do this in. That way, you can keep a close eye, but not become obsessed with it.
- Whenever possible, spend a few bucks to get the best-rated equipment. If you are on a tight budget, you can pick less sophisticated equipment that is made by a recognized manufacturer.

In my experience, I have found that most folks are kind and will share their experiences with you. Ultimately, going off the grid is a big decision. So, the folks that do so end up developing a good sense of community. So, they are willing to help others who are looking to go off the grid. If you want to learn some specifics about installing your solar power system you can reach them online and ask questions! Hope you were enjoying reading the book.

Thank you for reading!